Twenty-First Century Issues

Twenty-First Century Issues

Perspective on Climate Change, Pandemic, and Digital Transformation

Edited by
Ashraf Esmail, Brandon Hamann, Edward Agbai, Henrietta M. Okoro, and Sylvester Udeorah

HAMILTON BOOKS
Lanham • Boulder • New York • London

Published by Hamilton Books
An imprint of The Rowman & Littlefield Publishing Group, Inc.
4501 Forbes Boulevard, Suite 200, Lanham, Maryland 20706
www.rowman.com

86-90 Paul Street, London EC2A 4NE

British Library Cataloguing in Publication Information Available

Library of Congress Cataloging-in-Publication Data Available

ISBN 978-0-7618-7388-4 (cloth)
ISBN 978-0-7618-7389-1 (electronic)

Contents

Acknowledgement

We express our profound gratitude to the President of the World Association of Academic Doctors (WAAD), Dr. Henrietta Okoro, the Officials of WAAD, and the planning committee of the 2022 book project for conceiving the idea of writing the book - *21st Century Issues: Perspective of Pandemic, Climate Change, and Digital Transformation*, which highlights the Issues of the 20th Century and spotlights the challenges and solutions for a brighter future globally.

We are especially grateful to the numerous contributors. This book would not have been possible without the constant push from the leader of the Book Project Committee, Dr. Edward Agbai, who made concerted efforts to ensure that the team for the *21st Century Issues: Perspective of Pandemic, Climate Change, and Digital Transformation* book was on track to meet timelines. Finally, our sincere appreciation goes to the Reviewers, Editors, and the Publishers of Hamilton Books.

Sylvester Udeorah, PhD

Preface

The idea for this book came from the World Association for Academic Doctors (WAAD) meeting held on Sunday, the 12th of December 2021. The book project is to actively engage members in collegiate scholarly work beyond publishing articles and a forum for bonding and teamwork. WAAD has a journal that publishes articles annually with members participating in the publication. The book project will harness the potential of members interested in writing a book or participating in one as co-authors but are not provided with the opportunity. The book "Living in the 21st Century World: Perspective on Climate Change, Pandemic, and Digital Transformation" signals a new direction for WAAD and its leadership.

The 20th Century produced the industrial revolution that propelled industrial growth through the steam engine and ironically led to increased global warming; it also had the 1918 Spanish flu that devastated much of the world. The 20th Century ended with the birth of the internet, signaling the start of the 4th industrial revelation.

The 21st Century entering its third decade, witnessed an exponential rise in the possibilities of digital technology through digital transformation, the threat to global peace and stability through the frightening projections of climate change and a pandemic (Corona virus pandemic) that held the world captive for two years. What lessons from the history of the last Century? What comparison can we make? And what is the way forward for humanity? This book seeks to provide insight into these questions and more.

Stakeholders are desirous of preserving the planet for the next generation while enjoying the benefits of digital technology in a healthy and prosperous world. This book makes them aware of the realities we face and the possible solutions available. It also provides them with the support and insight to uncover what they truly want and the encouragement and tools to create a better world.

The contribution to the manuscript was developed mainly by members of WAAD and other experts on the subject matter. It provides perspectives, stories, book recommendations, key messages, and famous quotes. With tables, diagrams, and illustrations included the audience for this book will be people of age, with a particular focus on business and political leaders influencing policies and decisions of government, including multilateral organizations. Young and emerging entrepreneurs, not for profit organizations, development partners, climate change gatekeepers and academia will also be interested in this book.

The book competes in trendy categories: self-development and personal transformation, as the thematic area covered are trending and pertinent to our present reality as a people. It covers climate change that global leaders, both political and business leaders, are worried about, digital transformation that is changing the way we do business and the coronavirus pandemic—a once-in-a-lifetime pandemic with both health and economic consequences. Contributing authors from different backgrounds encourage readers to act and save our planet.

Because contributing authors live in different parts of the globe, having real-time meetings was challenging. Still, with the aid of technology, we deployed asynchronous ways of collecting members' contributions to produce this unique masterpiece that takes a retrospective look at our past hundred years to proffer solutions relevant to the challenges we face as a people in the twenty-first Century.

After reading this book, readers can determine our experiences of the 20th Century, identify similarities with what we currently face, learn lessons from these experiences, and have foresight into the future based on our activities today. Because the book will provide linkages between the activities of the last Century and the present day, project into the future with verifiable data to make reasonable assumptions, educate others on the need to protect the environment and ultimately become part of the solution rather than the problem.

Section 1

THE TWENTIETH CENTURY

A RETROSPECTIVE VIEW

Chapter 1

The Spanish Flu and its Effects on Population Growth

The Aftermath

By Jessica L. Burke, Lisa A. Eargle & Ashraf Esmail

INTRODUCTION

Pandemics existed globally as early as the Justinian Plague, which affected Egypt, the Eastern Roman Empire, and surrounding areas in years 541 and year 543 (Cunha & Cunha, 2008). Since the Justinian Plague, several other pandemics have spread throughout history, including the Black Plague, which affected Asia and Europe until the 19th century (Zietz & Dunkelberg, 2004), several cholera pandemics throughout 1800s and 1900s (Faruque et al., 1998), the Russian Flu that began in 1889 (Taubenberger et al,, 2007), the Spanish Flu that occurred between 1918 and 1919 (Johnson & Mueller, 2002), the Asian Flu that spread globally between 1957-1959 (Kawaoka et al., 1989), the 2009 H1N1 pandemic (Easterbrook et al., 2011), and finally the COVID-19 pandemic of today. Pandemics can impact on the countries they invade; most notable is the loss of thousands to millions of people. There are also lasting impacts resulting from governmental restrictions and lockdowns on entire nations, communities, and individuals, including backlash from those who favor protecting individual rights over governmental interference, temporary and permanent closure of businesses, and adverse mental health outcomes (Aragona et al., 2020; Greyling et al., 2021; Shang et al., 2021).

The Spanish Flu was the most devastating pandemic in modern history and coinciding with World War I made the outbreak even more damaging in terms of population loss and how governments managed the pandemic (Little, 2020; Shapiro, 2020). The purpose of this chapter is to examine the origins and influence of the Spanish Flu pandemic. Specifically, this chapter aims to address the following goals: (1) Explain the origins of the Spanish

Flu, including the mode of transmission; (2) Explore the aftermath of the pandemic on population dynamics, such as fertility, mortality, and immigration trends on a global and national scale; (3) Discuss the response of various world governments to manage the spread of the disease while also enduring World War I; and (4) Examine the aftermath of the Spanish flu, including impacts on health and health care accessibility, and disease management.

WHAT IS THE SPANISH FLU?

The Spanish Flu was a type of influenza. Influenza is a very contagious disease that attacks the respiratory system of its victims. The virus is transmitted from person to person (or across animals or from animals to humans) by breathing in tiny respiratory droplets from an infected entity. It is one of the diseases that has repeatedly created pandemics throughout human history. During these pandemics, the number of illnesses and deaths from influenza is estimated to be 80 times higher than during non-pandemic periods. Of the three types of influenza viruses that infect humans – A, B, and C -- A and B are the deadliest. The Spanish Flu was an A variety, which tends to occur more often than the other two varieties (Kuszewski & Brydak, 2000), and to be the only variety that causes pandemics (Cleveland Clinic, 2021).

Origins of Spanish Flu: Pathogenetic and Geographic Origins

Influenza has been around for over 1,000 years in some form. Some of the earliest known writings about it came from Hippocrates and Livy back in 421BC. Historians have uncovered evidence that suggests that numerous influenza pandemics have occurred within the last 4 centuries. In the early 1890s, Bujwid and Pfeiffer identified an entity called Bacillus influenza. Later that decade, the "Asian Flu" emerged in Russia in 1889. It first spread slowly, then quickly spread westward to North and South America and southward to Australia by early 1890. The death toll was more than one percent.

Afterward, mortality rates due to influenza dropped until around 1915, when an influenza epidemic emerged in the United States. In 1917, there were many cases of bronchitis amongst English soldiers stationed in France during World War I (1914-1918). In 1918, many American soldiers stationed in northwestern France also contracted the disease. It was called the "Spanish Flu" because news about the spreading disease was published in Spanish newspapers, but the information about the disease was censored in other nations (Kuszewski & Brydak, 2000).

The pandemic virus emerged/evolved through a series of steps. The human H1 virus which had appeared before in the middle of the 1st decade of the

1900s, somehow obtained an avian neuraminidase (N1) and protein genes. The resulting disease then appeared in swine for a short period, before appearing in humans. While the H1N1 influenza impacted very young children and the elderly, it disproportionately impacted those aged 20 to 40 years. It is believed that those both younger than 20 and older than 40 years had previously been exposed to other H1 strains of influenza, which provided them some protection. However, those who were young adults had been previously exposed to an H3N8 infection in their childhoods (during the years of 1889 to 1900), which did not provide them with immunity against the H1N1 Spanish Flu (Worobey et al., 2014).

Doctors made diagnoses of the disease based on symptoms – high fever, dry cough. runny nose, head and body aches, exhaustion, and lack of appetite. Many who fell ill also developed complications, such as pneumonia. There were no medical tests available to clinically determine infection, nor were there any medications or treatments to alleviate severe symptoms (such as difficulty breathing). Recent estimates have suggested that approximately 500 million people may have been infected worldwide (about one-third of the world's population at the time), with 50 million deaths occurring. In the U.S. alone, approximately 675,000 deaths were attributable to the Spanish Flu (Cleveland Clinic, 2021).

Spanish Flu Timeline and Spread

According to the Africa Center for Strategic Studies (2020), the first wave of the pandemic (March to August 1918) started in the Midwestern region of the United States and spread across the U.S., to western Europe, to Brazil, to the Middle East and from there into India, to eastern Asia, and into New Zealand and Australia. The second wave (August to December 1918) of the pandemic had infections spreading further in Europe, back to and within the United States, to the Caribbean, and from Europe into western and southern Africa, from southern Africa to India, from the Arabian Peninsula to India, and into the south Pacific, New Zealand, and Australia. The third wave (January to June 1919) spread from New Zealand into the south Indian Ocean, back to Africa and to places like Malaysia and Indonesia, and also across the Atlantic into the Americas again.

The first wave was much milder than the second wave, which was responsible for the most deaths. In the second wave, the disease had changed, becoming a highly infectious and lethal version of the virus. Much of the diffusion of the disease came through seaports, where people and goods were being transported into or exported from. As the disease spread, whole villages and towns in some countries were severely impacted (Africa Center for Strategic Studies, 2020).

Sometimes a pandemic may seem to appear out of nowhere, especially when a disease is spreading rapidly, like the Spanish Flu did. However, from examining multiple different pandemics over time, scientists have been able to discern different stages of disease spread and progression in a pandemic.

Stages of an Influenza Pandemic

According to the World Health Organization (WHO) 2009, there are six stages in an influenza pandemic. These stages are:

- Stage 1. No influenza viruses circulating in animals have been found to have created infections in humans.
- Stage 2. An influenza virus that was circulating amongst animals has now been found infecting humans. The potential threat for a pandemic to eventually occur is now possible.
- Stage 3. An influenza virus has been found to create small pockets of infection among human populations. Human-to-human transmission may be occurring, but under limited circumstances such as direct, close contact between an infected person and an unprotected person. The level of human transmission has not occurred at the level necessary yet to create a pandemic.
- Stage 4. An influenza virus is found circulating amongst humans, in human-to-human transmission, to an extent that community-level outbreaks are now occurring. The risk of a pandemic has increased dramatically, but this does not mean that a pandemic is a definitive outcome.
- Stage 5. Human to human transmission of the virus is now occurring in two or more countries. This is a strong indication that a pandemic is likely to occur in the near future.
- Stage 6. Community-level outbreaks of the virus are now occurring in two or more regions of the world. A global pandemic has now emerged. During this period, there may be multiple waves of high levels of infection among humans.

At some point, it is assumed there will be a post-pandemic era. In the post-pandemic era, influenza infections will decrease to levels usually seen for seasonal influenza outbreaks. Moreover, the pandemic virus will act in a manner typically seen for seasonal influenza. Note there is no time frame provided for quickly a virus that becomes a pandemic will emerge, progress through each stage of this model, or enter into the post-pandemic era.

The WHO also notes that different nations may be undergoing different stages of the pandemic at the same time. Also, in the case of an influenza virus that causes mild and unspecific symptoms, there may be widespread circulation of the disease in human populations before it is recognized as an emerging pandemic. Hence, it could appear as if a pandemic virus has skipped stages during the course of its emergence (World Health Organization, 2009).

The Spanish Flu meets the criteria of a pandemic in that it was a new version of influenza not seen before, even though there had been previous influenza outbreaks. The Spanish Flu was a virus with avian elements, which appeared briefly in swine, before it began to appear in humans (Worobey et al., 2014). There were instances of a few cases, then outbreaks, which occurred in the different regions (Americas and Europe initially) of the world before spreading rapidly within nations and across nations (Kuszewski & Brydak, 2000).

Every pandemic, though, ultimately involves the transmission of a disease from one person to another. This transmission just needs to occur repeatedly across large populations in relatively short period of time to create the pandemic. Thus, it may be useful to incorporate multiple disease transmission models into the discussion, to better understand how the Spanish Flu pandemic emerged.

How Disease Transmission Occurs

One of the most basic models of disease transmission and progression is the SEIR (Susceptible, Exposed, Infected, and Recovered) model. The model argues that a person becomes at risk of contracting a disease that is circulating around in the population. At some point, the person becomes exposed to the disease, If the disease successfully infiltrates the person's body, the person has become infected and has the potential to transmit the disease to others. Finally, the person is able to fight off the disease, symptoms decline, and the person is recovered from the illness (Aggarwal et al., n.d.).

These four major steps in the disease transmission and progression process can be further broken down into more detail using other disease transmission and progression models. For example, the Chain of Infection model can be used to further understand the dynamics involved in the transmission of disease from a person, animal, or place, to a person (a new disease host). The Chain of Infection model posits that there are seven components involved in the transmission of a disease. These components are the disease agent, a reservoir, infectious materials, a portal of exit, a mode of transmission, a susceptible host, and a portal of entry. A disease agent can be a virus (such as influenza), bacteria (cholera), fungi (Valley Fever), protozoa (malaria), or helminths (worms). These disease agents are contained in infectious materials (such as feces, saliva, blood, soil, or water), that are located inside of a reservoir (such as a human, animal, field, or waterway).

A reservoir is a location where the disease agent lives, grows, and replicates itself.

In order for the disease agent to be transferred from one location to another, the infectious materials containing the disease agent must have a portal of exit from the reservoir. These can be the excretion of materials from a human or animal, or the stirring of soil or water in an environment. Once the materials containing the agent have left the reservoir, depending on the type of disease involved, disease agents can be transferred elsewhere through different modes of transmission. This can occur through direct contact (such as touching or consuming infectious materials), through a vector (such as a mosquito or fly), droplets (from a sneeze or cough), or if a disease can be airborne (move in the wind or through heating/cooling systems).

The disease agent is then transmitted to a susceptible host, a person or other entity that is at risk of being infected. This occurs when the disease agent travels into a portal of entry in the susceptible host. This portal of entry can be the eyes, nose, mouth, blood stream, genitals, or skin. Once the disease agent has gained entry, a new reservoir or host for the disease has been created (Centers for Disease Control, 2012).

A third model, the Generalized Stages of Infection model, can be used to explain the progression of disease once it is inside of a person. Those Generalized Stages are entry, colonization, incubation, prodromal symptoms, invasive period, decline of infection, and convalescence. In the entry stage, a person is exposed to an infectious disease through some kind of contact with disease agents. In the colonization stage, the disease attaches itself to a person and begins to replicate. In the incubation stage the disease begins to spread throughout the person's body. A person begins to recognize signs such as fever and fatigue that indicate he/she has an infection. In the invasive period, secondary infections may emerge as well as tissue or organ damage. In the decline of infection stage, the symptoms of disease begin to decrease as the disease is being eradicated from the body. Finally, the person enters into a period of convalescence, where recovery from the disease in nearing completion (Hendrix, 2011).

A similar, but more detailed explanation of steps in the infection process is presented by Janeway et al. (2001). They specifically focus on the influenza virus and the immune system response to the virus. In step 1, a new person (susceptible host) is exposed to a disease through contact with infectious particles from a diseased person (current host). The ability of the disease to infect the new person depends on the prevalence, means of contact (mode of transmission), and strength and steadiness (stability) of the disease. Contact is made through an epithelial (body tissue) surface, which can be the skin, mucous membranes, or blood stream.

Next, after the disease has infiltrated the body, it sets up a site of infection in the new host. This requires the disease attaching itself to the body tissue of the new host and then colonizing it or piercing the tissue and replicating itself within the tissue. If the host's innate immune system mounts a suitable defense, the host can fight off and eradicate the disease. If the host's immune system is unable to eradicate the disease, then the disease spreads from the original site of infection, or issues toxins that spread to other parts of the body. The spread of the disease or its toxins occurs through the lymphatic system or blood stream.

As the spread of the disease occurs, the host shows symptoms of disease as the body develops an adaptive immune system response. The adaptive response involves antigen-specific T cells and antibody secreting B cells. These cells move through the lymphatic and blood systems to the sites of infection, where they launch attacks against the disease.

The T and B cells work to clear the body of the infection and its residues. After clearing the infection, the adaptive immune response system prevents a reinfection. This protection against reinfection may or may not be a total protection and, moreover, may or may not last over a prolonged period. Sometimes, even after a person has recovered from the infection, significant tissue or organ damage has been left behind by the infection. In some cases, however, the infection can go latent in the body and can reappear at a later time if the immune defense systems are weakened (Janeway et al., 2001)

Demographic Impacts of the Spanish Flu

There are several demographic impacts that a pandemic can have on a population. If the disease is easily transmittable and deadly, mortality rates will be high. With high mortality rates, life expectancy for a population may decline. It can also affect marital and fertility decisions, as people may postpone these decisions to a later and safer time. Finally, it also has the potential to impact migration decisions, as potential migrants may not want to migrate to an area experiencing disease, and destination populations may not be welcoming of newcomers that they know little about.

In the following sections, each of these demographic impacts are discussed further.

Mortality Totals and Rates by Nation

Death totals and rates varied greatly around the world. In terms of total deaths, India was the hardest hit with over 18 million deaths. However, with a total population of 305.6 million, the death rate in India was not the highest at .1 per 1000 population. Indonesia, for example, had 1.5 million deaths out of a population of 49.3 million, producing a death rate of 30.4 per 1000 population. China experienced approximately 4 million deaths out of a population of 472 million, with death rates estimated between 8 and 20 per 1000 population. Japan had 388,000 deaths, for a death rate of 7.

For the African region, the highest death counts were located in Nigeria (455,000 deaths in a population of 18.6 million) and South Africa (300,000 deaths in a population of 6.7 million). These nations also had high death rates (24.5 and 44.3 deaths per 1000 population). Kenya reported 150,000 deaths of out a population of 2.5 million, with a death rate of 57.8 per 1000 population. The highest death rate reported occurred in Cameroon, with 445 deaths per 1000 population. In general, Africa was one of the hardest hit areas of the world.

For the region of the Americas, the United States had the highest death total (675,000), but with a population of 103,2 million, the death rate was 6.5 deaths per 1000 population. The highest death rates were observed in Guatemala (39.2 deaths per 1000 population) and in Mexico (20 deaths per 1000 population).

For the European region, the highest death total is located in Russia/USSR, with approximately 450,000 deaths, but the death rate was 2.5 deaths per 1000 population. Italy had 390,000 deaths, with a death rate of 10.7 deaths per 1000 population. The highest death rates reported were Hungary (12.7 deaths per 1000 population) and Spain (12.3 deaths per 1000 population). France had a death rate of 7.3 and England had a death rate of 5.8 deaths per 1000 population (Niall et al., 2002).

Impacts on Life Expectancy

Estimates as high as 50 million deaths worldwide have been attributed to the Spanish Flu. In comparison, World War I killed 16 million from 1915-1918 (National Archives and Records Administration, n.d.) With this many deaths occurring, a precipitous drop in life expectancy occurred. In the U.S., life expectancy for men dropped from 52.5 years in 1915 to 36.6 years in 1918, before rebounding to 53.5 years in 1919. For women, life expectancy fell from 56.8 years in 1915 to 42.2 years in 1918. Before rebounding to 56.0 years in 1919. Those who had a tuberculosis infection were also more likely to succumb to influenza (Noymer & Gerianne, 2000).

Impacts on Marriage

According to the Association for Psychological Science, not much is known about pandemic impacts on rates of marriage and divorce. However, from patterns observed in other disasters, there are indications that a decline in marriage rates will occur during a pandemic and then rebound immediately in the year following. It is hypothesized that disasters (and pandemics) are external stressors on a romantic relationship, making it harder for couples to agree on issues and to be more critical of one another. However, divorce rates during a pandemic also decline (Association for Psychological Science, 2020).

Impacts on Fertility

Boberg-Fazio et al. (2021) found a short-lived baby boom in the United Kingdom from 1919-1919. In 1918, the birth rate was 18.3 births per 1000 population. In 1919, this increased to 23 births per 1000 population. However, this is also the immediate aftermath of WWI. In countries neutral like Sweden, birth increases were also experienced. Afterwards, though the general decline in fertility that was already underway continued into the mid-1940s.

Impacts on Immigration

In times of disease outbreaks, immigrant groups are often blamed for bringing the disease into the country. During the Spanish Flu, a major influx of immigrants from southern and eastern Europe into the United States continued. This large wave of immigration began in the 1890s and continued until the mid-1920s. In the years prior to the pandemic, some blamed the immigrants for the disease of polio. With the advent of the flu pandemic, some blamed the Germans for the disease (a country at odds with the U.S. during WWI). Since so many people who had immigrated to the U.S. or their children served in WWI, the animosity was directed abroad and not internally within the country (Kraut, 2010).

Response of World Governments to Manage the Spread of the Disease

The responses to manage and stop the spread of the Spanish Flu varied among different government bodies across the world. Responses ranged from censorship to denial, to social distancing, wearing masks, and improving personal hygiene practices, to no response at all. WWI also overshadowed responses to the Spanish Flu where the war effort was deemed more important than stopping the spread of a deadly flu (BBC News, 2020; Dayoff, 2020; Little, 2020). Some nations did not institute health measures to help eradicate the Spanish Flu until a later time. As a result, thousands of people died over a short period of time in many cities across the United States and Europe, as was the case in Philadelphia, Pennsylvania with over 12,000 deaths in approximately 4 weeks (Little, 2020) and during the deadliest wave of the flu, 1200 people were dying in Paris, France a week (Rettman, 2020). Censorship: delayed, ineffective, or no preventative measures to halt the spread of the flu; and denial and resistance created a perfect storm that brought disease and death in its wake on a global level (Dayoff, 2020; Little, 2020; Rettman, 2020).

Censorship in the U.S. and Europe

In 1918, U.S. Congress signed President Woodrow Wilson's Sedition Act into law. The Sedition Act made it illegal to say or publish anything negative that would go against the war or the country. Once the deadly flu outbreak occurred in the fall of 1918, the press, the government, and the military minimized the severity of the illness as it was believed that such news would hinder the war effort and produce a demoralizing effect on the country (Dayoff, 2020; Little, 2020). As a result, false news reports abounded. According to Spanish Flu expert John M. Barry, health officials on a federal and local level claimed that the Spanish Flu was an ordinary flu virus (Dayoff, 2020). Essentially, the war abroad took precedence over the thousands of people dying from the flu at home. Even though a reported 195,000 Americans died in one month (October 1918), the dire situation did not raise any cause of concern among U.S leaders (Shapiro, 2020). By the end of 1918, a reported 45,000 U.S. Army soldiers had perished from the flu, but that also did not raise any concern (Little, 2020). President Wilson never acknowledged the flu in public, and the hardest hit states and cities did not receive any federal assistance. Major newspapers in New York, Boston, and Philadelphia treated the Spanish Flu nonchalantly, generally reserving a softened one-column article on the front page. Moreover, *The Boston Globe* printed a headline that reported flu cases were declining, but such an assertion was flawed because the reporter compared that day's death toll with a day earlier in the week with a higher death rate. However, that same newspaper printed a graph that displayed how deaths were increasing yet the graph was buried on page nine. *The New York Times* buried stories on the flu's spread across the country on the last page of the paper, after the classifieds (Shapiro, 2020). Barry (2004) argued that due to the censorship by the government and the press, the news created a level of distrust among the public because the reports did not match the experiences of those who were sick or lost a loved one to the illness. Amid censorship, denial and mistrust, there was no consistency in implementing any health or prevention measures, and such measures were introduced, and mostly debated and contested well into the deadliest weeks of the pandemic (Shapiro, 2020).

In similar fashion to the U.S., most of Europe was fighting in WWI, resulting in the same censorship, denial, and a lack of real effort to eradicate the Spanish Flu. The mass casualties of war desensitized many Europeans; thus, the pandemic did not create mass hysteria among citizens. Moreover, one

British academic well-versed in the 1918 pandemic suggests that there were no known joint efforts among the U.S. and the European allies to implement any type of international intervention against the flu due to the threat of the war (Rettman, 2020). The news was censored in the European nations involved in the war, and these countries had stronger wartime censorship laws compared to the U.S. given their longer involvement (Little, 2020; Martini et al., 2019). Britain executed the Defence of the Realm Act or Dora, which considered any discussion of the Spanish Flu as a threat to nation morale during the war (Rettman, 2020). Thus, any news story that would cover how dangerous the flu was, or the death toll would not be published in newspapers as such information would hurt the country. During the spring and early summer of 1918, British news outlets claimed the Spanish flu was mild and was not a serious threat, and such sentiment continued by European newspapers during the deadlier second wave. For instance, Britain suppressed any news covering the Prime Minister's illness, who had contracted the flu while on a trip during the second wave (Little, 2020). The Royal Society of Medicine in Britain suggested for people to stay home if showing symptoms and to avoid large gatherings as early as July 1918, but the Government subjugated the information (BBC, 2020). The Italian Interior Minister denied that the flu was spreading in the country despite the increasing number of deaths. Censorship extended beyond any news that reported the spread of number of fatalities, but also included any comments or information that did not align with the nation's narrative of the illness. Although censorship laws were presented to protect morale among citizens, the laws also served to mask the devastating impact of the flu from enemy nations. The Spanish Flu claimed the lives of thousands of soldiers in Europe and such information could have played into war strategy (Little, 2020; Martini et al., 2019).

Public Health Measures in the U.S. and Europe

Although the war and censorship greatly constrained efforts to minimize the impact of the Spanish Flu, physicians were not required to report flu cases to health authorities further complicating matters (Spinney, 2017). Nonetheless, there were some public health measures implemented in the U.S. and Europe, particularly as the war was ending. Yet, some state and local governments and health officials rejected to implement any measures or enforce measures to stop the spread of the flu (Little, 2020). In addition to the war, there was public resistance to measures to help stop the spread of the flu, and public health authorities in both continents were not prepared to manage a pandemic as severe as the Spanish Flu. There was also variation in how different states implemented and enforced public health measures. The Committee of the American Public Health Association implemented policies to limit large gatherings, including closing saloons, theaters, dance halls, and bar nonessential meetings and funerals. Furthermore, the committee further suggested that stores and factories alternate their hours to reduce the number of people in a building at any one time and given the unsanitary nature of public transportation (e.g., street cars), people were encouraged to walk to work (The Public Health Response). Yet, the closure of public schools was not widely supported in the U.S. at the time (The Public Health Response); however, some schools did close by October of 1918 (Strochlic & Champine, 2020). Military bases enforced social distancing and prevented gatherings. Other measures targeted hygiene, and the committee called for educational initiatives, such as teaching the public the advantages of hand washing, stopping the use of shared eating utensils and cups, and teaching how to properly dispose nasal discharges. Posters to educate those about the dangers of the flu and how to reduce the spread of the disease were issued by some public health departments. Moreover, disinfection and sterilization were other important factors to combat the flu, particularly in hospitals. Various state level health departments also ordered those with the flu to quarantine until symptoms dissipated (The Public Health Response).

Some areas of the U.S. enforced mask laws. Masks were made of triangular shaped, folded gauze that covered the mouth, nose, and chin. The purpose of the mask was to prevent the spread of droplets from the nose and mouth, but it also prevented any infectious droplets to reach the hands and prevented any infection from the hands to reach the mouth (The Public Health Response). However, some of the public did not want to follow mask mandates. A health officer shot three people when one of them refused to wear a mask in San Francisco. Other areas of the U.S. imposed fines and jail sentences for individuals refusing to follow mask laws (Strochlic & Champine, 2020). Citizens also resisted large gathering restrictions and closures. The public also resisted against the prohibition of large gatherings and closures. For instance, the city of Philadelphia urged its citizens against coughing, sneezing, and spitting in public and physicians tried to tell the media of the dangers of public gatherings, but the city newspaper would not print the warnings. Then, despite physicians' pleas to the city's public health officials to cancel the event, the city hosted the "Liberty Loan March" parade where 200,000 people participated. As a result, Philadelphia experienced the highest peak death rate in the U.S. (Little, 2020; Strochlic & Champine, 2020). Citizens resisted other public health initiatives and mandates to slow the spread of the flu. Across the U.S., banning public gatherings showed difficult, particularly curbing religious gatherings (BBC, 2020).

Although New York City had the lowest death rate and was the first to initiate preventive measures, many business owners pressured the city's health commissioner to keep businesses open. New York City was in a unique position as the city had experience fighting tuberculosis and applied the same health measures to the Spanish Flu pandemic (BBC, 2020; Strochlic & Champine, 2020). Finally, St. Louis, Missouri implemented social distance guidelines within 48 hours of the first reported case, and their residents generally practiced social distancing. As a result, St. Louis was able to contain the virus and deaths more successfully than other cities. Thus, some places in the U.S. did initiate measures that worked to a degree, and people were not as resistant compared to other areas (National Institutes of Health, 2007; Parker, 2021).

The flu spread fast among European nations and in the U.S. According to Martini et al. (2019), war refugees and women participating in the workforce given men were away at war contributed to the spread of the flu, along with returning soldiers at the end of 1918. Moreover, the flu spread rapidly among those using public transportation, those working in factories, and troop transports and military camps (BBC, 2020). European nations involved in the war engaged in censorship and the press did not report on the severity of the flu. Nevertheless, there were some health measures implemented to halt the spread of the disease.

The following preventative measures in Europe included, but are not limited to: Public meeting places closed, church sermons were restricted to a maximum of five minutes, the number of people allowed on public transportation was limited, sanitation and disinfection of streets and public places regularly occurred, free soap and clean water was provided to the poor, food inspection and regulation was initiated, laws against public spitting were enacted, funerals were prohibited, voluntary or sometimes mandatory isolation and quarantine of citizens, and crowds and gatherings were banned (Martini et al., 2019). Preventative measures varied greatly across Europe, as they did across the states. For example, measures were weaker in Britain than in the U.S. As opposed to closing theaters and other places of entertainment, music hall performances were shortened, and several intermissions were implemented to allow for ventilation time. Yet, on the other hand, Europe was more responsive to closing schools. Britain closed public elementary schools and France initiated a policy that any student with symptoms was to stay home, along with their siblings regardless of symptoms. Also, other countries in Europe, particularly those not involved in the war, were much more likely to implement preventative measures, such as closing all places where people could gather (Public Health Response). There were some health campaigns advising citizens on how to avoid the Spanish Flu, including to walk more instead of public transportation, forcing one to sneeze morning and night, and washing the inside of one's nose. While some preventative measures were employed throughout Europe and the U.S., there was no government-mandated lockdown for an entire nation or geographical area to stop the spread of the flu (BBC, 2020).

Responses to the Spanish Flu: Asia and Africa

The Spanish Flu impacted the world beyond the U.S. and Europe. Not under any censorship laws, China acknowledged the flu pandemic in early June 1918 as the Shanghai daily newspaper reported the number flu cases and published a severe outbreak alert. Officials also closed public places to curb the spread of the flu. Chinese citizens used herbal medicines to try to treat the disease and sprayed their roofs with limewater to prevent the flu (Zhang et al., 2019). Moreover, it has been suggested that traditional Chinese medicine may have helped slow the spread and lessen the severity of the disease (Farley, 2020).

China, Japan, and India were all involved in WWI in some capacity, but China did not send troops to combat (Boissoneault, 2017). However, this was not the case for Japan and India. Japanese troops were struck with the Spanish Flu in November of 1918, and it is estimated that six to eight percent of the Japanese army perished from the disease (Farley, 2020). As for Japanese citizens, the first wave of the Spanish Flu hit Japan in September 1918 at a textile factory, where large numbers of female employees fell ill. Through October, hundreds had died and many essential employees, such as public transportation drivers and telephone operators, were sick with the flu resulting in service disruptions. By the second wave in 1919, six preventative measures were implemented in Japan: Maintain distance from those who are sick, evade crowded areas, use a mask to cover the mouth and nose, gargle often, wash hands often, and to be mindful of the needs of both children and elderly (Kuchikomi, 2020). India, a British colony and thus a part of the British Empire during WWI, contributed heavily to the war effort for Britain. India sent approximately 1.5 million soldiers to fight for the allied powers and gave supplies and money to Britain. As for the Spanish Flu, it spread through India in a devastating fashion. However, since India was under Britain's rule, officials denied the severity of the disease while millions were infected and succumbed to the flu (Farley, 2020).

Africa was also impacted by the Spanish Flu, but countries implemented vast public health measures, including stressing the importance of sharing information regarding the disease throughout the continent. Alerts regarding the flu were disseminated early on through newspapers and telegraphs, and such means of communication was used to warn medical authorities when ships with known flu outbreaks were arriving at port cities. Ships were also inspected, and mandatory quarantines were imposed on those entering any port. People were urged to cooperate with all health measures to halt the spread of the virus. The main measures used across African countries was social distancing and quarantines. All public places were closed, including schools, churches, markets, and roads. Oftentimes, schools and churches served as makeshift emergency hospitals. Mandates banning large gatherings were instituted as well. Two countries, Zanzibar, and Nyasaland imposed a mandatory quarantine on its citizens and contact tracing was used to impede and eradicate the flu. Officials in Lagos also attempted to monitor and contact trace infected people, but the system became overwhelmed due to the vast number of cases. In addition, health workers and volunteers worked at emergency hospitals, soup kitchens, and depositories that gave food and medicine to those in need. These same volunteers disinfected places where outbreaks occurred. Generally, Africans were receptive of preventative measures, and many volunteered their time regardless of race or social class (Africa Center for Strategic Studies, 2020).

Pandemic Aftermath and Worldwide Impacts

Armistice, or the end of WWI, happened on November 11th, 1918, and many historians suggest that the deadly second wave of the Spanish Flu prompted the end of the war given the loss of soldiers on both sides (Little, 2020). After the third wave in early 1919, the world began a long road to recovery after war and flu devastation. As a result of the 1918-1919 deadly pandemic, nations changed how health matters were viewed and managed. Perceptions that illness was the fault of the individual because they were poor or did not work hard enough transformed because the Spanish Flu crossed racial and class lines – the upper echelon of society was not spared. Accessibility to health care was at the forefront of changes in both the U.S. and Europe. Reporting of contagious illnesses became mandatory and standardized, and research on public health became an interest about a decade after the flu. After the worst pandemic in global history, the consensus of most everyone was the following: Health is a basic human right, not a privilege. Such sentiments, along with the massive loss of life, spurred improvements in public health and health care around the world (Spinney, 2017).

Inequalities and Healthcare

Eugenics was the prevailing view of health officials and the wealthy during the Spanish Flu outbreak. The poor and racial minorities were deemed inferior human beings and due to this perceived inferiority, they were naturally prone to disease. Moreover, the wealthy blamed the poor and racial minorities if they were to fall ill and die, simply arguing that such people lacked the motivation to attain a healthier life. Social factors or fundamental causes of disease, such as poverty, poor diet, overcrowding, were not considered. Thus, public health measures were designed to protect the wealthy from acquiring diseases from poor, working-class individuals and other second-rate classes of people (Spinney, 2017; Spinney, 2020). Overall, the poor and working class were impacted more by the Spanish Flu in terms of illness rates and number of deaths. For instance, African Americans were 1.5 times more likely to die from the Spanish Flu in Florida and tended to be poorer with less access to health care than whites (Gabriel, 2020). However, with the Spanish Flu, the public health measures designed to protect the wealthy failed as the elite were not unaffected from illness or death. As a result, health officials and the wealthy could no longer use eugenics to justify why an impoverished person died from the flu. Also, blaming an individual for falling ill with an infectious disease was no longer practical as was treating such diseases on an individual level – whole populations needed to be treated. As a result, eugenics fell by the wayside after the 1918-1919 pandemic, and the approach to health care by authorities and physicians across the globe was the need for more inclusivity and accessibility to all people (Spinney, 2017; Spinney, 2020).

Despite this change in viewpoint, the COVID-19 Pandemic unveiled a similar pattern of disease and death 100 years later in the U.S. – poverty and one's racial or ethnic background is still a risk factor for negative outcomes from communicable disease pandemics (Patel et al., 2020). Using patient data from 14 different Mass General Brigham System sites in the Boston area, Wasfy and colleagues (2021) find that communities with a median household income of $70,000 or less were three times more likely to test positive for COVID-19 than communities with higher median household incomes. Also, areas with higher levels of unemployment (more than 5%) were twice as likely to acquire COVID-19 than communities with higher employment rates. In this study, Hispanics were more likely to test positive for COVID-19 at 17.2% than African Americans (11.9%) or whites (5.6%). When it comes to infectious disease management, there are lessons learned, but lessons that still need to be initiated by public health officials and the government. Social inequalities still exist today as they had in 1918, which indicates a deep-rooted structural impetus that contributes to poorer health outcomes, including the spread of infectious diseases, among the impoverished and racial minorities.

Healthcare Delivery

Prior to the Spanish Flu pandemic, most people did not have access to physicians and general healthcare. After the pandemic, however, governments on a global scale began to understand the importance of accessible healthcare. Furthermore, the Spanish Flu being a worldwide health crisis, unveiled the need for a global joint effort involving public health officials and physicians to combat future pandemics. During the Spanish Flu pandemic in conjunction with WWI censorship laws, public health authorities were excluded from political meetings, particularly at the federal level, and were left imploring for assistance from local health departments or other departments often unsuccessfully. Hence, an international bureau for fighting epidemics, which was the impetus for the World Health Organization (WHO), was formed in Austria in 1919. Beginning in the 1920's as a direct result of the Spanish Flu pandemic, most of Europe and Russia chose to establish free healthcare for all citizens, or socialized medicine funded by a state backed insurance system. The United States did not employ socialized medicine, instead opting for an employer-based insurance system. Moreover, alternative medicine grew exponentially in the United States and beyond as more people were turning away from science post pandemic, given there was no cure developed to eradicate the flu. China's reaction post pandemic was the opposite. China embraced science as means to fight future pandemics by implementing improved public health measures, increased and improved disease surveillance and reporting, and more healthcare data collection. Regardless of the method utilized and despite the flaws of both private and socialized medicine, the Spanish Flu pandemic led to the acknowledgement of world governments to increase medical accessibility for all individuals and the need to create an international effort or organization to prevent and manage global pandemics (Spinney, 2017; Spinney, 2020; Whiting, 2020).

Managing Pandemics Post Spanish Flu

The importance of systematically reporting disease numbers across entire populations and the use of epidemiology to inform public health measures were two related outcomes of the Spanish Flu pandemic. The world envisioned a different type of medicine in the 1920's, one that would include three major elements: Biology, experiments, and sociology. Thus, modern day public health as a discipline that relied on systematic, scientific data on disease outbreaks along with using the data to design health safety measures was a direct result from the Spanish Flu pandemic. However, for this vision of public health to materialize, data on the population's health was needed. As a result, the field of epidemiology was officially recognized as a science (Spinney 2017; Spinney 2020). Epidemiology is "the study of the distribution and determinants of health-related states or events in specified populations, and the application of this study to the control of health problems" (Last, 2001:61). The two disciplines of public health and epidemiology work together to not only inform the public about health concerns, but to also protect people. Due to this acceptance of epidemiology, all U.S. states were reporting to a national disease reporting system by 1925, and only a decade later the first national health survey was implemented to collect data on the health status of U.S. residents (Spinney, 2017; Spinney, 2020). Infectious disease surveillance was another measure used to track and monitor the spread of disease. This tactic, however, did not work to slow the spread of the Spanish Flu given wartime censorship and the subsequent delay in implementing health measures. By the Asian Flu outbreak in 1957, surveillance was more successfully utilized. Specifically, the WHO initiated a worldwide surveillance network to trace and provide global warning when this flu began to spread (Tognotti, 2013).

Hence, advancements in the views of infectious diseases, healthcare accessibility and public health measures have resulted from the devastating impact of the Spanish Flu. Yet it is worthy to note that many of the same tactics used to alleviate the spread of the Spanish Flu have been implemented in pandemics worldwide since although not to the same degree (Huremovic, 2019). Largely, responses to pandemics have been virtually the same since the Black Death that began in 1347 (Patterson et al., 2021). For example, quarantining populations has a rich history as a tool used to combat infectious disease - used as early as the Black Death (Tognotti, 2013) and as early as 1647 in the U.S. colonies in response to smallpox and yellow fever outbreaks (Grateful American Foundation). In 2014, the Center for Disease Control (CDC) initiated a 21-day voluntary quarantine in response to the Ebola outbreak in West Africa. States and local governments can also impose their own quarantines, and many implemented a mandatory 21-day quarantine for those suspected of Ebola exposure (Dahl et al., 2016). More recently in 2020, quarantine was also used globally in response to COVID-19. In the U.S., the federal government, and many state and local governments, initiated mandates to "stay home" and only essential workers were to report to their jobs (AJMC, 2021). Quarantines are currently regulated by section 361 of the Public Health Service Act (42 U.S. Code § 264) and gives power to the U.S. Secretary of Health and Human Services to institute any necessary measure to avert the spread of infectious diseases in the U.S., whether from foreign countries or between states. The U.S. Secretary of Health and Human Services has granted the CDC the authority to issue quarantine or isolation mandates, in addition to examining, detaining, and releasing individuals exposed to an infectious disease (Center for Disease Control, 2021).

Social distancing is another measure that was used to curb the Spanish Flu pandemic as previously discussed earlier in this chapter. This practice involves restricting groups, prohibiting mass gatherings, closing schools and other public places, including theaters, restaurants, and churches, enacting travel restrictions both domestically and internationally, and maintaining physical distance from other people (Huremovic, 2019; Persons, 2020). The purpose of social distancing is to inhibit the transmission of communicable diseases from person-to-person to control the spread of disease and thereby lessening the peak number of infections or what is known as "flatten the curve" (Persons, 2020). Social distancing guidelines are still currently utilized as a preventative measure to manage pandemics globally (Humremovic, 2019), as was the case with the COVID-19 Pandemic (Persons, 2020).

While the use of social distancing as an effective method to eradicate a pandemic is debated (Barry, 2007), there is one consequence that has been widely discussed because of COVID-19 – a consequence not widely discussed as an issue during the Spanish Flu pandemic, but an issue, nonetheless. According to Huremovic (2019), social distancing results in isolation, ranging from physical separation via barriers (e.g., windows) or protective gear (e.g., masks) to social separation from loved ones, and uncertainty where those who are sick are uncertain about recovery and survival and those who are not sick are uncertain if they will fall victim to the virus, and economic uncertainty due to lockdowns or mandated closures of businesses. The mental health consequences of both quarantine and social distancing are profound on a global scale. Historical research indicates a positive relationship between mental illness, suicide, and the Spanish Flu in the U.S. Psychiatric hospitalizations increased during the years following the pandemic, and those that survived the flu reported a host of psychological symptoms from depression to sleep difficulties (Mamelund, 2003). Britain had a more stoic and fatalistic response to the death and devastation caused by both the war and the pandemic, where the British viewed death as more normalized given the sheer fatalities from the war (Bayly, 2020).

With more awareness and acceptance of mental health issues today, COVID-19 has revealed the severity of mental illness in relation to social distancing and quarantine during a pandemic. Research shows an increase in post-traumatic stress symptoms, depressive symptoms, anxiety, and lower psychological well-being among individuals with confirmed COVID-19 infection, the public or individuals not infected, those with pre-existing psychological disorders, and healthcare workers (Vindegaard & Benros, 2020). Talevi and colleagues (2022) suggest that a psychological crisis intervention model needs to be initiated globally for those affected by pandemics, including implementing online mental health services that would be available worldwide.

Advancements in Medicine: Vaccinations

Medical science further advanced after the 1918-1919 pandemic. During the Spanish Flu outbreak, scientists believed it was caused by a bacterium, but by the Asian Flu pandemic, they understood that influenza was caused by various viral strains. Even in 1918, many health officials wanted to develop a vaccine to stop the spread of the disease, but the disease was an unknown in terms of its true cause and the science was lacking at the time to develop an effective vaccine (Schwartz, 2018). Spanish Flu vaccines were utilized, but targeted bacterial strains rendering them ineffective (Barry, 2004). Over a decade later in 1936, two influenza A vaccines were developed (Plotkin & Plotkin, 2018). Influenza pandemics since, including the Asian Flu and the novel H1N1 flu outbreak in 2009 resulted in mass vaccination campaigns and international public health discussions on the development, production, and dissemination of flu vaccines. Annual flu vaccines, or commonly known as "flu shots" have gained increased support during the second half of the 20th century (Bauer, 1985; Neustadt & Fineberg,1982). Today, scientists continue researching to develop a vaccine that could cover most, if not all, flu strains, and continue researching to develop vaccines to prevent Ebola, Zika, small-pox, and anthrax (Paules et al., 2017; Schwartz, 2018) and vaccines are still being developed in response to managing Covid-19. The WHO suggests that the COVID-19 pandemic has led to an unprecedented interest in the development, regulation, and safety of vaccinations (The World Health Organization, 2022). Not unlike the resistance against health measures that occurred during the Spanish Flu pandemic (Little, 2020; The Public Health Response), people also lack confidence in health and safety measures, including vaccines during the COVID-19 pandemic despite advancements in medicine (World Health Organization, 2022).

CONCLUSION

This chapter examined the origins of the Spanish Flu, the population impacts of the pandemic, the response to the outbreak to curb the spread of the flu, and the aftermath of the Spanish Flu. As the most devastating pandemic in modern world history with an estimated 50 million total deaths worldwide (Cleveland Clinic, 2021; Little, 2020; Shapiro, 2020), there were subsequent impacts on life expectancy, fertility, and immigration. Efforts to contain the illness where overshadowed by WWI, and such efforts were generally ineffective given the inconsistencies and delays in implementation (Dayoff, 2020; Little, 2020). Improvements in healthcare occurred as a result of the pandemic, but socioeconomic inequalities persist today. Pandemics impact those who are poor and racial minorities more than other groups, and the Spanish Flu was no exception, and a little over 100 years later the COVID-19 pandemic reminded the world that such inequalities in health outcomes remain (Patel et al., 2020). Moreover, public health measures initiated during the Spanish Flu, including mask mandates, social distancing, quarantine, and sanitation are the same methods utilized before the Spanish Flu pandemic and afterwards (Huremovic, 2019; The Public Health Response). The effectiveness of these measures is debated (Barry, 2007), and there are negative outcomes related to the public health measures of pandemics that are deadlier than the pandemic itself. For instance, food production and basic health care and health needs are often disrupted, and chronic and mental illnesses are associated with major economic downturns. Hence, researchers suggest that policymakers and public health authorities need to weigh the benefits with the risks of quarantine and lockdowns during a pandemic, considering the direct impact of the illness and the costs of public health measures on communities and individuals (Brenner, 2020; Patterson et al., 2021). Despite the advancements resulting from the Spanish Flu pandemic (Spinney, 2017; Spinney, 2020), lessons still need to be learned to manage and fully understand the effects of current and future pandemics (Patterson et al., 2021).

REFERENCES

Africa Center for Strategic Studies. (2020, May 13). *Lessons from the 1918-1919 Spanish Flu Pandemic in Africa.* https://africacenter.org/spotlight/lessons-1918-1919-spanish-flu-africa/

Aggarwal, A., Hwang, A., & Jain, N. (n.d.). Productionizing a COVID-19 System Dynamics Model for the Indian Government. https://web.stanford.edu/~ashishg/msande433/India_Model/india_model/report.html

AJMC. (2021, January 1). A timeline of COVID-19 developments in 2020. The Center for Biosimilars. https://www.ajmc.com/view/a-timeline-of-covid19-developments-in-2020

Aragona, M., Barbato, A., Cavani, A., Costanzo, G., & Mirisola, C. (2020). Negative impacts of COVID-19 lockdown on mental health service access and follow-up adherence for immigrants and individuals in socio-economic difficulties. *Public health, 186*, 52–56. https://doi.org/10.1016/j.puhe.2020.06.055

Association for Psychological Science. (2020). Pandemic effects on marriage and relationships. https://www.newswise.com/coronavirus/pandemic-effects-on-marriage-and-relationships

Barry, J. M. (2004). The great influenza: The story of the deadliest pandemic in history. New York, NY: Penguin Books.

Barry, J. M. (2007). Comment on: Non-pharmaceutical interventions implemented during the 1918-1919 influenza pandemic. *JAMA*, 298(19), 2260-2261.

Bauer, D. J. (1985). A history of the discovery and clinical application of antiviral drugs. *Br. Med. Bull.*, 41(4), 309-314.

Bayly, M. (2020, October 28). *Fatalism and an absence of public grief: How British society dealt with the 1918*. LSE British Politics and Policy. https://blogs.lse.ac.uk/politicsandpolicy/public-memory-1918-flu/

BBC News. (2020, May 10). *Coronavirus: How they tried to curb Spanish Flu Pandemic in 1918*. https://www.bbc.com/news/in-pictures-52564371

Boberg-Fazio, N., Ivets, M., Karlsson, M., & Nilsson, T. (2021). Disease and fertility: Evidence from the 1918–19 influenza pandemic in Sweden. *Economics and*

Human Biology, Volume 53. https://www.sciencedirect.com/science/article/pii/S1570677X21000447

Boissoneault, L. (2017, August 17). *The surprisingly important role China played in WWI.* Smithsonian Magazine. https://www.smithsonianmag.com/history/surprisingly-important-role-china-played-world-war-i-180964532/

Brenner, M. H., (2020). Will there be an epidemic of corollary illness linked to a COVID-19-related recession? *Am. J. Public Health*, 110:974-975. doi: 10.2105/AJPH.2020.305724

Cunha, C. B., Cunha, B. A. (2008). Great plagues of the past and remaining questions. (D. Raoult, & M. Drancourt, Eds.). *Paleomicrobiology*. Springer, Berlin, Heidelberg. https://doi.org/10.1007/978-3-540-75855-6_1

Centers for Disease Control. (2012). The Principles of Epidemiology in Public Health and Practice, Third Edition On-line. Atlanta: CDC.

Centers for Disease Control. (2021). *Legal authorities for isolation and quarantine.* https://www.cdc.gov/quarantine/aboutlawsregulationsquarantineisolation.html

Cleveland Clinic. (2021). Spanish Flu webpage. https://my.clevelandclinic.org/health/diseases/21777-spanish-flu

Dahl, B. A., Kinzer, M. H., & Raghunathan, P. L. (2016). CDC's response to the 2014-2016 Ebola epidemic-Guinea, Liberia, and Sierra Leone. MMWR Suppl., 65(Suppl-3), 12-20. https://doi.org/10.15585/mmwr.su6503a3.

Dayoff, K. (2020, March 20). *Government censorship made the 1918 Spanish Flu even worse.* The Baltimore Sun (Carroll County Times). https://www.baltimoresun.com/maryland/carroll/lifestyles/cc-lt-dayhoff-032220-20200320-wmahqlv3drbghpv2hziglq4jny-story.html

Easterbrook, J. D., Kash, J. C., Sheng, Z., Qi, L., Gao, J., Kilbourne, E. D., Eichelberger, M. C., & Taubenberger, J. K. (2011). Immunization with 1976 swine H1N1- or 2009 pandemic H1N1-inactivated vaccines protects mice from a lethal 1918 influenza infection. *Influenza and Other Respiratory Viruses*, 5(3), 198–205.

Farley, R. (2020, March 25). *Remembering the Spanish flu in Asia.* The Diplomat. https://thediplomat.com/2020/03/remembering-the-spanish-flu-in-asia/

Faruque, S. M., Albert, M. J., & Mekalanos, J. J. (1998). Epidemiology, genetics, and ecology of toxigenic Vibrio cholerae. *Microbiol Mol Biol Rev.*, 62(4):1301-1314. doi: 10.1128/MMBR.62.4.1301-1314.1998. PMID: 9841673; PMCID: PMC98947.

Gabriel, J. (2020, February 20). *Understanding the socioeconomic dimensions of deadly pandemics.* UCF Today. https://www.ucf.edu/news/student-research-understanding-the-socioeconomic-dimensions-of-deadly-pandemics/

Grateful American Foundation. (2022). *Why did Congress pass the federal quarantine legislation?* https://gratefulamericanfoundation.com/11532/

Greyling, T., Rossouw, S., & Adhikari, T. (2021). The good, the bad and the ugly of lockdowns during Covid-19. *PLoS ONE* 16(1): e0245546. https://doi.org/10.1371/journal.pone.0245546

Hendrix, J. (2011). *Infection and disease.* http://science.kennesaw.edu/~jhendrix/ahmicro/handouts/infection.ppt&rct=j&q=stagesoftranamissionofcommunicabledisease

Huremovic, D. (2019). Chapter 8: Social distancing, quarantine, and isolation. (D. Huremovic, Ed.). Springer. https://doi.org/10.1007/978-3-030-15346-5_8

Janeway, C., A., Jr., Travers, P., Walport, M. (2001). Immunology: The Immune System in Health and Disease. 5th edition. New York: Garland Science.

Johnson, N. P. A. S., & Mueller, J. (2002). Updating the accounts: Global mortality of the 1918-1920 "Spanish" influenza pandemic. *Bulletin of the History of Medicine, 76*(1), 105–115. http://www.jstor.org/stable/44446153

Kawaoka, Y., Krauss, S., & Webster, R. G. (1989) Avian-to-human transmission of the PB1 gene of influenza A viruses in the 1957 and 1968 pandemics. *J Virol.*, 63(11):4603-46088. doi: 10.1128/JVI.63.11.4603-4608.1989. PMID: 2795713; PMCID: PMC251093.

Kraut, A. M. (2010). Immigration, Ethnicity, and the Pandemic. Public Health Reports Supplement.

Kuchikomi. (2020, May 6). How the Spanish flu of 1918-20 ravaged Japan. Japan Today. https://japantoday.com/category/features/kuchikomi/how-the-spanish-flu-of-1918-20-ravaged-japan

Kuszewski, K., & Brydak, L. (2000). "The epidemiology and history of influenza. *Biomed & Pharmocother,* 54: 188-95.

Last, J. M. (2001). Dictionary of Epidemiology. 4th ed. New York: Oxford University Press.

Little, B. (2020, May 26). *As the 1918 flu emerged, cover-up and denial helped it spread.* History.com. https://www.history.com/news/1918-pandemic-spanish-flu-censorship

Mamelund, S. E. (2003). Effects of the Spanish influenza pandemic of 1918-19 on later life mortality of Norwegian cohorts born about 1900. Working paper. https://www.researchgate.net/publication/5097223_Effects_of_the_Spanish_Influenza_Pandemic_of_1918-19_on_Later_Life_Mortality_of_Norwegian_Cohorts_Born_About_1900.

Martini, M., Gazzaniga, V., Bragazzi, N. L., & Barberis, I. (2019). The Spanish influenza pandemic: A lesson from history 100 years after 1918. *J. Prev. Med.*, 60, E64-E67. https://doi.org/10.15167/2421-4248/jpmh2019.60.1.1205

National Institutes of Health. (2007, April 2). *Rapid response was crucial to containing the 1918 flu pandemic.* https://www.nih.gov/news-events/news-releases/rapid-response-was-crucial-containing-1918-flu-pandemic

Neustadt, R. E., & Fineberg, H. (1982). The epidemic that never was: Policy-making and the Swine Flu Scare. New York, NY: Random House.

Niall, P., Johnson, A. S., & Mueller, J. (2002). Updating the Accounts: Global Mortality of the 1918-1920 "Spanish" Influenza Pandemic. *Bulletin of Historical Medicine*, 76: 105-115.

Noymer, A., & Gerianne, M. (2000). The 1918 influenza epidemic's effects on sex differentials in mortality in the United States. *Population Development Review*, 26 (3): 565-581.

Parker, T. A. (2021). The U.S. response to the Spanish flu of 1918 and COVID-19. *H-SC Journal of Sciences*, X. http://sciencejournal.hsc.edu/

Patel, J. A., Nielsen, F. B. H., Badiani, A. A., Assi, S., Unadkat, V. A., & Patel, B. (2020). Poverty, inequality, and COVID-19: The forgotten vulnerable. Public Health, 183, 110-111. doi: 10.1016/j.puhe.2020.05.006

Patterson, G. E., McIntyre, K. M., Clough, H. E., & Rushton, J. (2021). Societal impacts of pandemics: Comparing COVID-19 with history to focus our response. Frontiers in Public Health, 9(630449). doi: 10.3389/fpubh.2021.630449

Paules, C. I., Marston, H. D., Eisinger, R. W., Baltimore, D., & Fauci, A.S. (2017). The pathway to a universal influenza vaccine. *Immunity*, 47(4), 599-603. doi: 10.1016/j.immuni.2017.09.007.

Persons, T. (2020, May). *Science & tech spotlight: Social distancing during pandemics*. GAO: Science, Technology Assessment, and Analytics. https://www.gao.gov/assets/gao-20-545sp.pdf

Plotkin, S. L., & Plotkin, S. A. (2018). A short history of vaccination. (S. A. Plotkin, W. A. Orenstein, P. A. Offit, K. M. Edwards, Eds.). Plotkin's Vaccines. 7th Ed. Philadelphia, PA: Elsevier.

Rettman, A. (2020, March 30). *How Europe coped with pandemic 100 years ago*. EUobserver. https://euobserver.com/health-and-society/147918

Schwartz, J. L. (2018). The Spanish flu, epidemics, and the turn to biomedical responses. *AJPH*, 108(11), 1455-1458.

Shang Y, Li H., & Zhang, R. (2021). Effects of pandemic outbreak on economies: Evidence from business history context. *Frontiers in Public Health*, 9(632043). *doi:* 10.3389/fpubh.2021.632043

Shapiro, W. (2020, March 31). *How America's newspapers covered up a pandemic*. The New Republic (TNR). https://newrepublic.com/article/157094/americas-newspapers-covered-pandemic

Spinney, L. (2017, September 27). *How the 1918 flu pandemic revolutionized public health*. The Smithsonian Magazine. https://www.smithsonianmag.com/history/how-1918-flu-pandemic-revolutionized-public-health-180965025/

Spinney, L. (2020, March 7). *The world changed its approach to health after the 1918 flu. Will it after the COVID-19 outbreak?* Time. https://time.com/5797629/health-1918-flu-epidemic/

Strochlic, N., & Champine, R. D. (2020, March 27). *How some cities "flattened the curve' during the 1918 flu pandemic*. National Geographic. https://www.nationalgeographic.com/history/article/how-cities-flattened-curve-1918-spanish-flu-pandemic-coronavirus

Taubenberger, J. K., Hultin, J. V., & Morens, D. M. (2007). Discovery and characterization of the 1918 pandemic influenza virus in historical context. *Antivir Ther*, 14(4 Pt B), 581-591. PMID: 17944266; PMCID: PMC2391305.

Talevi, D., Socci, V., Carai, M., Carnaghi, G., Faleri, S., Trebbi, E., Di Bernardo, A., Capelli, F., & Pacitti, F. (2022). Mental health outcomes of the COVID-19 pandemic. *Riv Psichiatr*, 55(3), 137-144.

The Public Health Response. (n.d.). The 1918 influenza pandemic response. Stanford University. https://virus.stanford.edu/uda/fluresponse.html

Tognotti, E. (2013). Lessons from the history of quarantine, from plague to influenza A. Emerg Infect Dis., 19(2), 254-259. https://doi.org/10.3201/eid1902.120312.

Vindegaard, N. & Benros, M. E. (2020). COVID-19 pandemic and mental health consequences: Systematic review of the current evidence. *Brain, Behavior, and Immunity*, 89, 531-542. https://doi.org/10.1016/j.bbi.2020.05.048

Wasfy, J. H., Hidrue, M. K., Rockett, H., Cafiero-Fonseca, E. T., Ferris, T. G., & del Carmen, M. G. (2021). Socioeconomic and demographic characteristics of both inpatients and outpatients with positive testing for SARS-CoV-2. *J. Gen. Intern. Med.*, 36(8), 2522-2524. doi: 10.1007/s11606-021-06919-x

Whiting, K. (2020, April). *A science journalist explains how the Spanish flu changed the world.* World Economic Forum. https://www.weforum.org/agenda/2020/04/covid-19-how-spanish-flu-changed-world.

World Health Organization. (2009). Pandemic Influenza Preparedness and Response: A WHO Guidance Document. Geneva: World Health Organization.

World Health Organization. (2022). *Statement for healthcare professionals: How COVID-19 vaccines are regulated for safety and effectiveness (revised March 2022).* International Coalition of Medicines Regulatory Authorities and World Health Organization. https://www.who.int/news/item/17-05-2022-statement-for-healthcare-professionals-how-covid-19-vaccines-are-regulated-for-safety-and-effectiveness

Worobey, M., Han, G., & Rambaut, A. (2014). A synchronized global sweep of the internal genes of modern avian influenza virus. *Nature*, 508 (7495):254-7.

Zhang, R., Dong X., Wang D., Feng L., Zhou L., Ren R., Greene C., Song Y., Millman A. J., Azziz-Baumgartner E., & Feng Z. (2019). One hundred years of influenza since the 1918 pandemic - Is China prepared today? *China CDC Wkly.*, 1(4), 56-61. PMID: 34594605; PMCID: PMC8422199.

ZIetz, B.P., Dunkelberg H. (2004). The history of the plague and the research on the causative agent Yersinia pestis. *Int J Hyg Environ Health.*, 207(2):165-78. doi: 10.1078/1438-4639-00259. PMID: 15031959; PMCID: PMC7128933.

Chapter 2

The Rise of Global Citizens

By Ghosh Thinnakkakath

INTRODUCTION

More people are becoming global citizens as they think beyond borders and cultures. In the 21st century, global citizenship has become increasingly important, thanks to increased opportunities for international employment and the widespread use of information and communication technologies. Anyone with an internet connection can communicate instantly with anyone, anywhere in the world, via any media. Digital technologies and cloud computing have entirely transformed how we do business. Getting to know different cultures and interfacing with international residents is a fantastic experience. However, it is sad to think that millions of people worldwide lack the necessities of life. Nationality dynamics are changing, and citizens are becoming more egalitarian and diverse as global citizens become more prevalent. Embracing our global citizenship will help to overcome many barriers between us. Global citizenship is precious, and we can all work together to prevent unnecessary wars, restore peace, and alleviate poverty by working together. One earth, one citizen, and a motto of peaceful living without poverty can bring about a better world for all.

CIVILIZATION AND MODERNIZATION

Approximately 5500 years ago, the first civilization appeared, and modernization occurred in the 18th century (He, 2014). Historians disagree about when globalization began; some point to 1492 and 1498, but actual globalization started in the early 19th century (O'Rourke & Williamson, 2002). Humanity needs cultural and religious diversity to be healthy and progress intellectually and spiritually. The social-religious reform movements created a national awakening among the masses. Religion can enhance human characteristics in some ways, but it isn't necessarily selective in what it enhances. Religion multiplies generosity if the leader is generous. A religion of humanity promotes peace, order, and discipline in society. It guides human behavior positively, leading to social prosperity and maintaining humankind's existence.

Social revolutions have happened in the past and will likely happen again based on history to bring about significant changes in social, economic, and political structures. Civilization has seen many social reformers throughout history. It's impossible to ignore the sacrifices made by those great social reformers in shaping up the more civilized human society we have today, even if it isn't perfect for everyone. There are many great reformers worldwide such as Mahatma Gandhi, Sri Narayana Guru, Harriet Tubman, and Florence Nightingale. A significant number of reformers helped shape the society and world we live in today. Sri Narayana Guru worked hard to eliminate India's caste system and discrimination with his principle of "One Caste, One God, and One Religion for All" (Olson, 2006). According to Hobson (2014), Harriet Tubman is the most famous African American woman in the world due to her effort to escape and rescue others from slavery. It is still a struggle to achieve equality and equity in the modern world.

Humanism is a system of thinking that emphasizes the value of humans rather than religious doctrines and faith. Humanism and secularism formed the basis for modern civilization, not religion. Humanity benefits from secularism, and it is not antagonistic to religions. Those who live in secular liberal democracies are happiest. Secular societies provide absolute religious freedom, where citizens can practice any religion they choose, with no religion privileged over others. Conflict and violence have been part of human history since the beginning. Their fighting becomes more intense and ferocious as they move from the domestic societal level to the national and international level. For societal control, secular materialism and spiritual humanism play crucial roles in winning people's hearts and minds. By creating a community of self-enforcing, self-monitoring individuals, they ensure the smooth function of the system and achieve cooperative progress (Sharif, 2004).

The issue of gender inequality has also been prevalent in the past, as evidenced by unequal wages at the workplace and sexual harassment that was considered to be part of social constructs. Different regulatory actions led to significant improvements in gender discrimination. During the 1992 Earth Summit, a green economy was envisaged for the sake of development and poverty eradication, according to which equal opportunity for all was the most important factor (Babugura, 2017).

GLOBALIZATION

In our old civilizations, we explored trade routes and promoted cultural exchanges to facilitate globalization. Globalization accelerated the movement of goods and services throughout the world. Cultural interaction between communities around the world is the main effect of globalization. After the second half of the 20th century, technological advancements contributed to globalization on a larger scale. In addition to economic and financial globalization, there is also a cultural, technological, and environmental component. It has benefited economies and enhanced cultural exchanges across the globe. One of the negative impacts of globalization is that income inequalities among people are widened. The number of billionaires has increased significantly, and 82% of all wealth produced in the last year went to the top 1% and none to the bottom 50% ("Reward work, not wealth", 2020). Globalization grew steadily due to technological advances in all spheres of life. Rapid developments in Information and Communication Technology have brought everyone closer together. The internet and social media have allowed people to communicate across the globe. In today's globalized economy, multinational corporations (MNCs) provide economic and employment opportunities worldwide. Nowadays, outsourcing from one organization to another has been moved overseas to save costs to survive in the market.

As part of globalization, offshore outsourcing negatively impacted jobs and manufacturing in developed countries. Shao and David (2007) state offshore outsourcing creates many challenges and controversies for businesses, government agencies, and IT workers. Many highly qualified and experienced IT workers lost their jobs in developed countries when companies started outsourcing heavily to foreign service providers. Recent economic anxiety has focused on the relationship between job losses, offshore outsourcing, and immigration (Hoffman, 2010). Deindustrialization and automation have contributed to the loss of millions of manufacturing jobs, which has significant electoral implications across advanced democracies (Baccini & Weymouth, 2021). Manufacturing and middle-income knowledge economy jobs have declined steadily in the United States (Arogyaswamy & Hunter, 2018). Globally, automation and the recent pandemic took another toll on jobs. Large technology companies are now getting cheap labor from other countries, job opportunities are disappearing, and young people are looking for good jobs in neighboring countries.

CULTURAL CONFLICTS AND LEADERSHIP ISSUES

Our identity is shaped by culture, and it plays a vital role in our everyday lives. The globalization of today has made cross-cultural communication easier. Employees can become bored, unhappy, and unproductive due to cultural differences. Ineffective communication and micromanagement can lead to conflict and tension. When people from different cultural backgrounds work together, they may have different beliefs and values. Liu et al. (2020) cited cultural misunderstandings as a potential source of risk for international joint ventures. As globalization accelerates, multicultural issues are also on the rise. Managing multicultural environments and understanding cultural diversity is crucial for success. Leaders face significant challenges in dealing with issues related to global leadership. Effective management of people from different cultures requires global leadership skills. Leaders must adapt to these changes to create a more culture-friendly working environment. Emotional intelligence plays a critical role in managing multiple cultures.

International trade has grown due to the increase in global interactions, as have exchanges of ideas, beliefs, and cultures. Interaction and integration are two critical economic processes associated with globalization's social and cultural aspects. Cultural intelligence and emotional intelligence are essential for global leaders to function effectively. An individual with cultural intelligence can understand another culture's actions to work more effectively. Self-awareness is fundamental to developing cultural intelligence. Crowne (2009) suggests that cultural intelligence differs from emotional intelligence in some ways and that cultural heritage influences emotional expression.

POPULATION EXPLOSION

The population explosion is another problem that needs to be resolved by providing adequate education and awareness of future issues. Population explosions have an enormous impact on the environment and the economy. With population growth rates reduced, sustainable development could be achieved. McNamara (1992) noted that it took thousands of years for the human population to grow. Over a million years passed before 1 billion people were reached at the beginning of the last century, and then the rate increased. A second billion was added in 130 years, a third in 30 years, and a fourth in 15 years. Globally, there are approximately 7 billion people. Over half of the world's population lives in Asia, while one-quarter lives in Africa. Global population growth has been attributed to high growth rates in Africa and Asia and declining mortality rates.

Several attempts have been made in the past to reduce population explosions, and it worked well. The population explosion is over, with fertility levels below the replacement rate in most advanced countries and rapidly declining in most developing countries (Siegel, 2012). Economic growth will be robust in the next half-century, especially in developing countries, and world wealth will rise dramatically. As a result, it will be easier to preserve the natural environment and avoid resource shortages. To protect our planet, we must raise awareness about the risk of population explosion, and the actions of global citizens can have a significant impact.

DIGITAL TRANSFORMATION

Information Technology (IT) plays a vital role in every aspect of our lives. With today's IT infrastructure, the world is more closely connected than ever, and we now live in a world where doctors can perform critical procedures from a distance. IT strategies are essential for businesses today. Information Technology enhances decision-making through its powerful computing speed, business intelligence reports, and Artificial Intelligence (AI). With Enterprise Resource Planning (ERP), policymakers can gather relevant information at their fingertips. Cloud computing enables organizations to store any amount of data and other required computer resources online and on an on-demand basis without owning a data center. Every industry is adopting digital technology to improve efficiency, improve business processes for optimal growth, gain better insights, innovate, and improve security. In business, artificial intelligence has a significant impact, and as noted by Bock et al. (2020), it is expected to bring more substantial changes like another industrial revolution.

Information sharing and other interests are now central features of social media. Everyone uses Google and other social media search engines to find news and additional information. The rise of social media usage has contributed to the decline of mainstream media usage globally (Park et al., 2020). The presence of smart mobile devices has become so ubiquitous that it is hard to imagine life without them. Piller et al. (2012) noted that social media has a tremendous impact, and co-creation with customers is rising, resulting in more efficient innovations. Utilizing social media and other smart devices excessively can affect job performance, and according to Yu et al. (2018), social media exhaustion can affect job performance.

GLOBAL PEACE AND COOPERATION

Many unnecessary battles have been fought in history that caused catastrophic and severe tensions. We all know that war is a destructive method that creates more tensions and chaos than it solves. Global peace and cooperation are essential for the survival of humans and other species on earth. Global warming and the next world war will have disastrous consequences in the future.

Human civilization may become extinct in the future due to fires generated by nuclear weapons or slow burning caused by global warming (Morgan, 2009). Peace can be ensured by preventing wars. The UNESCO charter states that war begins in the minds of men, so the defense of peace should also begin in the minds of men. Recent pandemics and concerns about climate change and global warming paved the way to explore life in space and other possible planets. Billings (2019) noted we are not yet ready, but maybe humans will be capable technologically and socially of living together peacefully on other planets in the distant future.

GLOBAL CHALLENGES/ISSUES

The 21st century offers all sorts of amenities that our elder generations could not have imagined. In just a few seconds, everyone on earth has access to the internet and can see and interact with each other. There are several global challenges, such as poverty, climate change, and pandemics. It is a wake-up call for everyone who has recently dealt with the unexpected effects of the recent pandemic. The pandemic forced everyone to stay at home to emphasize the importance of life and protect themselves from harm. If everyone believes in themselves and supports each other, they can all make a difference. There has been an increase in the development and research of drugs to prevent or mitigate future pandemics. The extreme poverty of people is a lingering issue that should be addressed as soon as possible.

Globalization brought us many benefits. The modern world is being transformed by globalization, and the process continues in every field. In the age of globalization, everyone can interact with anyone in a matter of seconds. There are many issues caused by globalization, such as cheap labor being exploited in underdeveloped and developing countries. Laborers in domestic manufacturing lose their jobs when countries outsource manufacturing to cheaper competitors overseas. In the United States, deindustrialization has occurred as most companies have moved their business to China and other countries. The growth of e-commerce businesses has caused most local retail businesses to close in the US, and we are already seeing many vacant business shops and manufacturing facilities.

In the history of humanity, globalization has been the most progressive force. We do not want one country's collapse over globalization, even though there are many benefits due to globalization. Globalization requires better management, and the recent pandemic COVID 19 accelerated the transformation of globalization. The exponential growth of digital technology paved the way for faster and more efficient business operations. COVID-19 is a revelation for all who want to create a more sustainable and more peaceful world through better management of globalization to ensure everyone lives without poverty and can prosper and achieve equality.

THE DANGERS OF NUCLEAR WAR

Two world wars have occurred in the past, and those were the most extensive and brutal conflicts in human history. The first world war, also known as the great war, lasted from 1914 to 1918 (Vinogradov, 2015). During the Second World War, which lasted from 1939 to 1945, millions of people died as a result of genocide, starvation, and other diseases. On 24 October 1945, the United Nations Organization was founded to promote international peace, human rights, technology, and economic development (Pinteală, 2018). As the principal organ of the United Nations, the Security Council is responsible for ensuring international peace and stability. Due to the Security Council's structure, interests, and priorities of its members, and staying away from critical global issues, the effectiveness of the United Nations has been questioned in recent years.

It is alarming to see the recent pandemic and the war between Russia and Ukraine. The current conflict could become dangerous if global peace organizations like the United Nations don't take any action to mediate and reduce fatalities. We need a world without nuclear and other destructive weapons. Conflicts between nuclear powers or accidental detonations, an irresponsible leader, or a terrorist act can lead to dangerous scenarios. Global citizens have a crucial role to play in achieving global peace. To promote peace and prosperity for everyone on earth, great leaders must be elected by their citizens everywhere. Peace will reduce the risk of nuclear conflict. The likelihood of a nuclear war is reduced when we reduce the chances of any war.

CLIMATE CHANGE AND GLOBAL WARMING

Global warming refers to the rise in global temperatures caused by a concentration of greenhouse gases in the atmosphere. On the other hand, climate change refers to a long-term shift in climate measurements. We are experiencing extreme droughts throughout the world because of climate change. As sea ice melts in the Arctic, sea levels rise, and coastal areas are at risk. Human life, ecosystems, and the environment are all impacted by climate change. Climate change is influenced both by natural as well as human factors. Global warming threatens all living beings unless nations reduce greenhouse gas emissions. Sustainable energy solutions are a reality and attainable. Utilizing renewable energy will reduce the effects of climate change. Life on earth is dependent upon a healthy planet.

COVID 19 PANDEMIC

There have been unexpected infectious diseases throughout history, some of which have spread worldwide as pandemics. COVID-19 is one of the worst pandemics in human history. Businesses suffered losses locally, nationally, and globally. The decrease in production impacts other industries and countries that supply parts, machinery, and services through global value chains (Court et al., 2021). Thousands of people lost their jobs. Due to outsourcing and cost-cutting, the production of essential products depends on multiple countries. This has had a detrimental effect on GDP and employment worldwide. In the findings of Brannen et al. (2020), Covid-19 will permanently decrease young people's confidence in political institutions and their leaders and their governments' healthcare policies.

Socioeconomic Inequality and Extreme Poverty

Although the world has changed a lot since the internet age, poverty worldwide remains a problem. Poverty can be caused by a lack of resources or control over resources, political issues, a lack of access to education, and geography. Extreme poverty and poverty aren't the same things. In general, poverty refers to an individual not being able to afford his basic needs. There is poverty everywhere, and even rich countries suffer from poverty and homelessness. In US cities, nearly 930,000 people lacked access to sanitary facilities, and 610,000 lacked access to basic water sources (Beng et al., 2020). Extreme poverty is a severe form of poverty that is bad for humanity. As Sengupta (2010) explains that extreme poverty is caused by a lack of income, human development, and social exclusion. Economic growth has lifted almost a billion people out of extreme poverty, but many still live in poverty. Sustainable and inclusive growth is necessary for eradicating poverty ("Why green development is so important," 2015).

Millions of people worldwide suffer from poverty, and the gap between the rich and the poor widens. Researchers and policymakers are paying increasing attention to the growing income gap between the rich and the poor (Larch & Philipp, 2020). Malnourished children are at a higher risk of dying from common illnesses. Nutrition-related causes of child deaths under five years of age account for 45 percent of all child deaths (Child mortality (under five years), 2022). Hunger is rooted in poverty, and extreme poverty poses a severe threat because millions lack access to clean water and food. Hunger is something we do not want to see. Rubenstein (2013) argues that eradicating global poverty is everyone's moral obligation. Many efforts are being made around the world to prevent poverty. In Target 1.1 of the Sustainable Development Goals, the international community is committed to eradicating extreme poverty by 2030 (Moon, 2017).

Sumner (2016) noted that poverty is shifting from low-income to middle-income countries, raising questions about the effectiveness of economic growth in reducing poverty. The United States of America is getting poorer when it comes to poverty. Based on the US Census Bureau's 2015 report, Mohammed et al. (2018) estimate that 43.1 million Americans live in poverty. According to Brady and Zachary (2020), approximately 5.2 to 7.2 million Americans are deeply poor, and 2.6 to 3.7 million are extremely poor. When homelessness is included, the numbers would be much higher.

GLOBAL CITIZENSHIP

The concept of global citizenship is all about sharing human values and responsibilities. By learning and sharing, we can secure a better future for everyone. The care and support of others go a long way toward relieving future worries. As Whiting et al. (2018) noted, the UN launched the Global Education First Initiative (GEFI) to accelerate progress towards universal access to education, fair and high-quality education, and the development of global citizenship. For all humans, the basic necessities of life are the same. Despite national sentiments or affinities, everyone can coexist on this beautiful planet. Everyone can contribute to the well-being of humanity in various ways. Making a positive global impact requires sensitively and effectively adapting successful strategies to different cultures. Global citizenship is a mindset and a lifelong journey. Global citizens can consider things from different perspectives to understand problems better.

Global citizens are aware of global issues and respect others' cultures and diversity. Any change we make locally can have a global impact since we all live on the same planet. Through social activism and engagement, global citizens can bring about positive change. As technologies advance, the world is more closely connected than before, but societies are still divided over silly things. The lack of emotional intelligence among political leaders and other critical personalities causes unnecessary chaos. We need global citizenship to promote peace and harmony and protect our planet. Interaction among people increases as people share their values and beliefs.

Importance of Global Citizenship

Globally, hunger is a severe issue, and it has spread from poor to developed nations. Many organizations are working hard to assist poor countries, but this is not enough to make a difference. We should all support a world without hunger and poverty. By leveraging the support of wealthy nations, community groups, and financially stable citizens, we can achieve this goal sooner rather than later. Page and Pande (2018) argue that economic growth and redistribution of resources to the poor are not sufficient to end extreme poverty by 2030. A governing body that helps the poor is accountable and able to meet their needs. Even though there have been numerous attempts to eradicate extreme poverty, the rise of a global citizen can expedite the process. Every citizen must rise to the challenge of eradicating extreme poverty and poverty worldwide. There is no limit to what we can accomplish when we all work together. Through their actions, global citizens can contribute to peace and harmony without war or discrimination.

Shiva and Shiva (2019) believe that 'earth democracy' is essential for peace and sustainability in the future. War sometimes breaks out without any apparent reason. As Hausken (2016) demonstrated in his study on the cost-benefit analysis of war, initiating a war can lead to economic, human, and political gains or losses. In the 20-year war on terror, $8 trillion has been spent, and 900,000 have died ("Costs of the 20-Year War on Terror: $8 trillion and 900,000 deaths," 2021). Peace will allow us to use these losses for worthy humanitarian causes, including poverty eradication, combating climate change, and ensuring global stability. According to Jett (2019), peacekeepers' objectives are complicated when faced with extreme violence, and they are ineffective when dealing with it. Global citizens can help achieve global peace through collective action.

Whether an economy is capitalist or socialist, policies such as allocation and ownership differ. The leading market capitalist economies are moving to a purer form of capitalism, but a pure market capitalist system is unlikely to exist in history (Rosser & Rosser, 2018, p. 42). People from different religious and cultural backgrounds often have different behavioral motivations and action logics in economic globalization since religion is at the heart of human culture (Yang & Li, 2018). To advance intellectually and spiritually, it is vital to preserving cultural and religious diversity. Many factors contribute to global hunger, including corruption, unemployment, war, political and religious persecution, and a lack of education. People from all over the world can interact through social media thanks to the rapid development and growth of the internet and information technology. Every global citizen supports whatever it takes to secure a significant change. Nobody on earth should go to bed hungry due to a lack of food. Agricultural technology is advancing, so enough food can be produced for everyone.

The earth is home to all living creatures, including plants, animals, and all components of nature. The threat of global warming is a serious issue we are all facing, and all citizens have a responsibility to protect the earth. It is now time for people to move from self-interest to a broader ecological perspective for safeguarding the earth and promoting common interests for the good of everyone. The One Earth, One Future, and One Citizen vision can help us understand global climate change, resolve poverty, and live in peace.

The concept of global citizenship is growing as people are thinking beyond their cultures and borders. A group of rising global citizens provides support and service to the needy. There is no need for separation when it comes to helping each other since we all live on the same earth. Despite cultural differences and geographical differences, everyone has the same basic needs. Global citizens think outside the box and contribute generously to humanity and the environment. In today's increasingly connected world, everyone is responsible for reaching out to needy citizens and providing support in whatever way they can.

The Rise of Global Citizens

The life cycle is not eternal, all things alive will eventually decay and die. People achieve different things at different times in their lives. Either you can make your life beautiful, or you can just pass through it. It is the purpose of life to live and let live. As human beings, we must recognize that to live a life of peace and freedom. We must extend that same right to others, and you must not interfere with the right of others to live their lives as they choose. Social living is possible when members feel a sense of brotherhood and communal harmony. Several global peace movements, including Globalcitizen .org and the "Live and Let Live" movement, are working towards a more peaceful world.

The concept of global citizenship doesn't mean people have to give up their roots or nationality to live in peace around the globe. Earth is home to all life and should be protected. Global citizens are stakeholders in the global community. They realize the importance of working toward a common goal, understanding cultural values and diversity in unity and coexistence at the local and global level through their actions for social justice for all. It should be possible for global citizens to make wise decisions that benefit all. The concept of global citizenship does not oppose nationalism, religion, or beliefs. A rising generation of global citizens is characterized by unity within diversity and understanding beyond borders.

Global Citizens play a crucial role, and we must raise public awareness. Globalization is erasing boundaries, and people are becoming more and more connected. A growing movement in education reform seeks to cultivate global citizenship and think beyond national identity (Dill, 2012). Universities have become keen to develop programs that develop students' knowledge of global issues, compassion, and cultural competency due to globalization's impact on the economy, environment, and society. Education institutions increasingly emphasize global citizenship as a learning outcome (Mason & Thier, 2018).

The Value of Being a Global Citizen

Global citizens can contribute many values to humanity through their small actions. A miracle can happen combined with a mission and social responsibility. A person who can afford local or global support or assistance can make a big difference. Global citizens will lend a helping hand to those in need and those who have no voice to raise their concerns. War and other unimportant issues are now costing the world more money than they should, affecting the poor and everyone on the planet. Global citizens are for peace, which can bring more money to the needy. Today, people interact with social media and other communication platforms differently than in the past. Citizens worldwide are becoming more aware of their culture and willing to help others. Global citizens are broad-minded, anxious, and ready to support others to ensure that everyone can coexist peacefully in the world. Every one of us deserves a peaceful, discrimination-free life in this short and sweet lifetime. Our future generations can enjoy life more by coexisting without any barriers we have endured over the years.

CONCLUSION

The globalization process began many years ago, and social and religious reform movements led to a national awakening among the masses. Social revolutions led to profound changes in social, economic, and political systems. The humanism and secularism we see today are the results of modern civilization. A secular state ensures absolute freedom of religion. The past was marked by gender inequality, discrimination, and cultural and religious intolerance. Innovations in information and communication technologies have radically changed globalization and the interaction of people. Learning about cultures from around the world can be a wonderful experience. Cloud computing and digital technologies have revolutionized the way we do business. The fact remains that many people live in extreme poverty and hunger, despite living in a high-tech world. Hunger and extreme poverty are unacceptable worldwide since modern agriculture can provide enough food for everyone. People worldwide can connect instantly without borders or cultural barriers through the internet and social media. Humanity needs global peace to live together peacefully.

For the well-being of humankind, people from all countries need to think beyond nationality and culture. Shiva & Shiva's idea of 'Earth democracy' (2019) is essential for peace and sustainability. Poverty, climate change, and pandemics have increased global citizenship and importance. We must eliminate weapons of mass destruction from the world. It can be dangerous to have a conflict between nuclear powers of an irresponsible leader or accident involving a bomb, or a terrorist act. Global citizens have a vital role to play in achieving global peace. Climate change is causing severe droughts around the world. With the melting of Arctic Sea ice, sea levels rise, and coastal areas are at risk. Climate change affects human life, ecosystems, and the environment. Although globalization has many benefits, we do not want one country to collapse over it.

A pandemic like Covid-19 can have devastating effects on human life and shut down the entire world. It is a wake-up call for those who have recently dealt with unexpected consequences. Local, national, and global businesses suffered losses. Production of essential products is dependent on several countries due to outsourcing and cost-cutting. As a result of these factors, the global economy and employment have been adversely affected by the Covid-19 pandemic.

As global citizens become more prevalent, the dynamics of nationalism are changing, and citizenship is becoming more equal and diverse. Being proud of one's global citizenship will resolve many barriers among people. The value of being a citizen of the world is immense, and everyone can contribute to preventing unnecessary wars, restoring peace, and alleviating poverty by working together. Nationality dynamics are changing, and citizens are becoming more egalitarian and diverse as global citizens become more prevalent. By embracing our global citizenship, we can overcome many barriers between us. Global citizenship is precious, and we can all work together to prevent unnecessary wars, restore peace, and alleviate poverty by working together. One earth, one citizen, and a motto of peaceful living without poverty can lead to a better world for all.

REFERENCES

Arogyaswamy, B., & Hunter, J. (2018). The Impact of Technology and Globalization on Employment and quity: The Need for a Vigorous Corporate Response. *Proceedings of the Northeast Business & Economics Association*, 22–27.

Babugura, A. (2017). *GENDER EQUALITY: A CORNERSTONE FOR A GREEN ECONOMY*. South African Institute of International Affairs. http://www.jstor.org/stable/resrep28351

Baccini, L., & Weymouth, S. (2021). Gone For Good: Deindustrialization, White Voter Backlash, and US Presidential Voting. *American Political Science Review, 115*(2), 550–567. https://doi.org/10.1017/S0003055421000022

Beng, D. C., Cumming, O., M.Sc, Nichols, D., M.P.H., & Brown, J., PhD. (2020). Water and Sanitation in Urban America, 2017–2019. *American Journal of Public Health, 110*(10), 1567-1572.

Billings, L. (2019). Colonizing other planets is a bad idea. *Futures, 110*, 44. http://dx.doi.org/10.1016/j.futures.2019.02.020

Bock, D. E., Wolter, J. S., & Ferrell, O. C. (2020). Artificial intelligence: disrupting what we know about services. *The Journal of Services Marketing, 34*(3), 317-334. http://dx.doi.org/10.1108/JSM-01-2019-0047

Brady, D., & Zachary, P. (2020). The Levels and Trends in Deep and Extreme Poverty in the United States, 1993–2016. *Demography, 57*(6), 2337-2360. http://dx.doi.org/10.1007/s13524-020-00924-1

Brannen, S., Ahmed, H., & Newton, H. (2020). *Covid-19 Reshapes the Future.* Center for Strategic and International Studies (CSIS). http://www.jstor.org/stable/resrep25198

Child mortality (under 5 years). (2022, January 28). WHO | World Health Organization. https://www.who.int/news-room/fact-sheets/detail/levels-and-trends-in-child-under-5-mortality-in-2020

Costs of the 20-year War on Terror: $8 trillion and 900,000 deaths. (2021). Brown University. https://www.brown.edu/news/2021-09-01/costsofwar

Court, C. D., Ferreira, J.-P., Hewings, G. J. D., & Lahr, M. L. (2021). Accounting for global value chains: rising global inequality in the wake of COVID-19? *International Review of Applied Economics, 35*(6), 813–831. https://doi.org/10.1080/02692171.2021.1912716

Crowne, K. A. (2009). The relationships among social intelligence, emotional intelligence, and cultural intelligence. *Organization Management Journal, 6*(3), 148-163. http://dx.doi.org/10.1057/omj.2009.20

Dill, J. S. (2012). The Moral Education of Global Citizens. *Society, 49*(6), 541-546. http://dx.doi.org/10.1007/s12115-012-9599-8

Hausken, K. (2016). Cost benefit analysis of war. *International Journal of Conflict Management, 27*(4), 454-469. http://dx.doi.org/10.1108/IJCMA-04-2015-0023

He, C. (2014). Human civilization and world modernization. Civilization and Modernization. https://doi.org/10.1142/9789814603522_0001

Hobson, J. (2014). Harriet Tubman: A Legacy of Resistance. *Meridians, 12*(2), 1–8. https://doi.org/10.2979/meridians.12.2.1

Hoffman, M. E. S. (2010). Job Losses and Perceptions of Globalization. *Journal of World Trade, 44*(5), 967–983.

Jett, D. (2019). Why Peacekeeping Fails. *Middle East Policy, 26*(1), 89-96. http://dx.doi.org/10.1111/mepo.12402

Larch, M., & Philipp, M. (2020). Mitigating the Gap Between the Rich and the Poor. Key Trends and Drivers of Redistribution. *Intereconomics, 55*(4), 245-255. http://dx.doi.org/10.1007/s10272-020-0909-x

Liu, J., Cui, Z., Feng, Y., Perera, S., & Han, J. (2020). Impact of culture differences on performance of international construction joint ventures: the moderating role of conflict management. *Engineering, Construction and Architectural Management, 27*(9), 2353-2377. http://dx.doi.org/10.1108/ECAM-02-2019-0111

Mason, D. P., & Thier, M. (2018). Study Abroad, Global Citizenship, and the Study of Nongovernmental Organizations. *Voluntas, 29*(2), 404-418. http://dx.doi.org/10.1007/s11266-017-9899-0

McNamara, R. S. (1992). The Population Explosion. *The Futurist, 26*, 9. https://www.proquest.com/magazines/population-explosion/docview/218601297/se-2?accountid=10378

Mohammed, I., Hosen, M., & Mohammad Ashraful, F. C. (2018). Does poverty lead to crime? Evidence from the United States of America. *International Journal of Social Economics, 45*(10), 1424-1438. http://dx.doi.org/10.1108/IJSE-04-2017-0167

Moon, G. (2017). Race to End Poverty? The Roles of Ethnicity and International Economic Law in the Eradication of Extreme Poverty by 2030. *Nordic Journal of International Law, 86*(3), 366. https://www.proquest.com/scholarly-journals/race-end-poverty-roles-ethnicity-international/docview/1984839644/se-2?accountid=10378

Morgan, D. R. (2009). World on fire: Two scenarios of the destruction of human civilization and possible extinction of the human race. *Futures, 41*(10), 683-693. https://doi.org/10.1016/j.futures.2009.07.011

Olson, C. (2006). Modern Hindu Thought: The Essential Texts. *The Journal of Asian Studies, 65*(2), 449-450. https://www.proquest.com/scholarly-journals/modern-hindu-thought-essential-texts/docview/230392622/se-2?accountid=10378

O'Rourke, K. H., & Williamson, J. G. (2002). When did globalisation begin? *European Review of Economic History, 6*(1), 23-50. https://www.proquest.com/scholarly-journals/when-did-globalisation-begin/docview/207332425/se-2

Page, L., & Pande, R. (2018). Ending Global Poverty: Why Money Isn't Enough. *The Journal of Economic Perspectives, 32*(4), 173-200. http://dx.doi.org/10.1257/jep.32.4.173

Park, S., Fisher, C., Flew, T., & Dulleck, U. (2020). Global Mistrust in News: The Impact of Social Media on Trust. *The International Journal on Media Management, 22*(2), 83-96. http://dx.doi.org/10.1080/14241277.2020.1799794

Piller, F., Vossen, A., & Ihl, C. (2012). From Social Media to Social Product Development: The Impact of Social Media on Co-Creation of Innovation. *Unternehmung : Swiss Journal of Business Research and Practice, 66*(1), 7-7–27. http://dx.doi.org/10.5771/0042-059X-2012-1-7

Pinteală, G. (2018). UNITED NATIONS ORGANIZATION. *Quaestus,* (13), 101-105. https://www.proquest.com/scholarly-journals/united-nations-organization/docview/2120792533/se-2

Reward work, not wealth. (2020, August 6). Oxfam International. https://www.oxfam.org/en/research/reward-work-not-wealth

Rosser, J. B., & Rosser, M. V. (2018). Comparative economics in a transforming world economy (3rd ed.). MIT Press.

Rubenstein, J. C. (2013). Pluralism about Global Poverty. *British Journal of Political Science, 43*(4), 775-797. http://dx.doi.org/10.1017/S0007123412000385

Sengupta, A. (2010). Human Rights and Extreme Poverty. *Economic and Political Weekly, 45*(17), 85–93. http://www.jstor.org/stable/25664389

Shao, B. B., & David, J. S. (2007). The impact of offshore outsourcing on IT workers in developed countries. *Communications of the ACM, 50*(2), 89-94. https://doi.org/10.1145/1216016.1216026

Sharif, M. (2004). Religious-Historical Perspective on Conflicts and Violence: Secular Materialism versus Spiritual Humanism. *The International Journal of Sociology and Social Policy, 24*(1), 56-85. http://dx.doi.org/10.1108/01443330410790966

Shiva, V., & Shiva, K. (2019). One earth, one humanity vs. the 1%.

Siegel, L. B. (2012). Fewer, Richer, Greener: The End of the Population Explosion and the Future for Investors. *Financial Analysts Journal, 68*(6), 20-37. https://www.proquest.com/scholarly-journals/fewer-richer-greener-end-population-explosion/docview/1352763026/se-2?accountid=10378

Sumner, A. (2016). Why Are Some People Poor? *The European Journal of Development Research, 28*(2), 130-142. http://dx.doi.org/10.1057/ejdr.2016.2

Vinogradov, V. (2015). World War I: A Historian's View. *Social Sciences, 46*(3), 33. https://www.proquest.com/scholarly-journals/world-war-i-historians-view/docview/1778693057/se-2?accountid=10378

Whiting, K., Konstantakos, L., Misiaszek, G., Simpson, E., & Carmona, L. G. (2018). Education for the Sustainable Global Citizen: What Can We Learn from Stoic Philosophy and Freirean Environmental Pedagogies? *Education Sciences, 8*(4). http://dx.doi.org/10.3390/educsci8040204

Why green development is so important. (2015, July 28). World Economic Forum. https://www.weforum.org/agenda/2015/07/why-green-development-is-so-important/

Yang, L., & Li, M. (2018). Cultural understanding of global Governance: A perspective of religious culture. *Advances in Applied Sociology*, 08(05), 359-365. https://doi.org/10.4236/aasoci.2018.85021

Yu, L., Cao, X., Liu, Z., & Wang, J. (2018). Excessive social media use at work: Exploring the effects of social media overload on job performance. *Information Technology & People, 31*(6), 1091-1112. http://dx.doi.org/10.1108/ITP-10-2016-0237

Chapter 3

The Growth of Technology in the 20th Century

By Priscilla Okunji, Dana Mashal, Caron Strong, and Gina Brown

INTRODUCTION

Defined as the application of scientific knowledge to the practical aims of human life, technology is sometimes phrased as the change and manipulation of the human environment. It would be extremely difficult to write about current technology without outlining the history of great technologies that virtually propelled the power of innovative contemporary experiences. With respect to the recent history of technology, however, one fact stands out clearly: despite the immense achievements of technology by the 1900s, the 20th century witnessed more advance over a wide range of activities than the whole of previously recorded history.

A massive number of new technologies have been developed in the last century. Technologies such as electricity, the incandescent light bulb, the automobile, and the phonograph, first developed at the end of the 19th century, were perfected, and universally deployed. The first airplane to take flight was invented in 1903, and by the end of the century, large airplanes such as the Boeing 777 and Airbus A330 flew thousands of miles in hours. The development of the television and computers caused immense changes in information sharing. A look at all the greatest and most practical innovations that came into existence throughout the past century is warranted.

In the United States (US), four-in-ten technological advances are cited as the greatest American achievements. Overall, the single most-mentioned success is the space program. In one study, almost one-in-five participants cited the exploration of space as the America's most remarkable feat. The 20th Century can be described as an extremely fruitful and innovative time for technology. Each decade from this century saw at least five to ten new inventions that helped to shape society and the technology that is used today. Hence, this chapter will reveal most technologies that were invented prior to the present 21st century. The technologies' significance will be discussed based on their gains, challenges, and overall impact on public health with decade-to-decade highlights from 1900–2000.

1900-1910

The years 1900-1910 jump-started this innovation century with the invention of the modern escalator by Charles Seeberger (Bellis, 2020). An Iowa native, Seeberger patented his design for the escalator. He built it in conjunction with the Otis Elevator Company, where he was employed at the time, and his design won first prize at the Paris Exhibition of 1900 (NIHF, 2022). Notably, Seeberger' s design for the modern escalator was adapted and redesigned from Jesse Reno's invention of the original escalator in the late 1800 (Bellis, 2019).

Another renowned innovation with high utility today was the air conditioner by Willis Carrier. Considered the father of air conditioning, Carrier invented the air conditioner in 1902, just one year after graduating with his engineering degree from Cornell University (Bellis, 2021). His original calculations are still what is used today in our modern air conditioning systems (Bellis, 2021). His invention, especially its ability to control temperature and humidity, helped to flourish many other industries (Bellis, 2021).

Radio technology also had impressive advancements in 1901, with the first successful radio transmission to a radio receiver (Bellis, 2020). Although by this time, radio technology had been in existence, it was not until 1901 that people were able to send transmissions through them. 1901 was the beginning of multiple different additions and variations of radio to make it what it is today (Bellis, 2019).

Healthcare discoveries and technology were also flourishing in the early 1900s. In 1902, Sir James Mackenzie, a London-based cardiologist, used his work and studies on the heart, pulses, and arrhythmias to invent the first-ever polygraph, which is still in use to this day (Moore, 2017). 1903 was also an informative and innovative year in medical technology with the invention of the first electrocardiograph (ECG) (National MagLab, 2014). It was invented by Willem Einthoven, a Dutch Physiologist (National MagLab, 2014). Einthoven is, in fact, the scientist who labeled the different waveforms P, Q, R, S, and T, which the science and the medical community still uses today (National MagLab, 2014). Einthoven's electrocardiograph was a 600-pound, five-man operated machine, which has significantly been improved upon later through the 20th century to become the lightweight 12-lead tabletop machine with interpretation capabilities commonly used today (National MagLab,

2014). Consumer access to iterations of the ECG is now available as various pocket-sized devices with Bluetooth capabilities.

Welcomed in with much fanfare was the Wright Brothers' creation of the airplane in 1903 (Bellis, 2020). They developed a motor and an aircraft able to hold the weight of the motor and flew this machine in one year (Bellis, 2019). Unfortunately, their flight only lasted 12 seconds, but in 1904 they were able to sustain flight for five minutes, and by 1908 they were operating passenger flights (Bellis, 2019). During this time, the French were working on inventing and flying the first helicopter. While Paul Cornu created the original design and prototype, Russian-American aviation engineer Igor Sikorsky's design in 1910 was built upon later to create the modern-day helicopter (Bellis, 2019).

By 1904, the chemistry and physics communities were advancing as well. John Ambrose Fleming announced his inventions of the electron tube, fondly named after himself 'the Fleming Valve' as well as the vacuum tube diode (Bellis, 2019). The purpose of these tubes was to regulate and control the movement of electrons and electricity and are still used today in televisions and monitors but are now supplemented by using plasma and LCD (Bellis, 2019). They were later improved upon in 1906 by Lee de Forest, who invented the triode, which aided in the magnification of electrons (Bellis, 2020). Also, in 1904, the electromagnetic waves were studied, and the primary version of the radar was invented by German engineer Christian Hulsmeyer (Trueman, 2015). The radar's counterpart, sonar, was invented two years later, in 1906 by Lewis Nixon (Bellis, 2020). Nixon's first purpose for the use of sonar was for the detection of icebergs, and its use flourished during World War I, when the need to detect submarines increased (Bellis, 2020).

1911-1920

The Model T automobile was invented by Henry Ford in 1908 and led to the mass production of automobiles. The Model A was the first car Ford invented under the Ford Motor Company in 1903. Although the first gasoline-powered car was developed in the late 1800s by Karl Friedrich Benz and Gottlieb Daimler, it was Ford that brought the car to many American households. American inventor Charles Ketting revolutionized driving the automobile in 1911 by creating the electrical ignition, which made it possible to start a car from inside the cabin and eliminated the use of the hand crank (Bellis, 2019).

Leading up to World War I in 1914, there was the invention of the military tank in 1912 and the gas mask in 1914 (Bellis, 2020). Pursuit planes and the timing mechanisms for machine guns mounted on those planes were also developed at that time (NAHF, 2020). During this decade, the radio industry saw many technological improvements. It began with the radio circuit in 1918 by Edwin H. Armstrong (Lessing, 2021). After serving in the army during

World War I, this American inventor improved radio signals by inventing the superheterodyne circuit, which helped to enhance the way of receiving, converting, and amplifying electromagnetic waves (Lessing, 2021). The flipflop circuit was invented by William Eccles and F.W. Jordan in 1918 (Alexander, 2019). These circuits are still in use today in data storage and transfer, also in registers, counters, and computers. (Alexander, 2019).

In the realm of household objects, this decade brought about the refrigerator (Bellis, 2020). In 1913, Alfred W. Wolf Jr invented the first refrigerator for domestic use. While it was not as successful, many other versions were based upon Wolf's invention and algorithms (National Academy of Science, 2022).

1921-1930

The medical community saw a jump in innovative technology in the 1920s with the inventions of the hearing aid in 1923 and the iron lung in 1927 (Bellis, 2020). The version of the modern-day hearing aid that was invented by Naval engineer Earl Hanson was derived from vacuum tube technology (Dahl, 2021). Hanson's breakthrough with hearing aids led to scientists and inventors perfecting the technology to what is sold today (Dahl, 2021). A few years later, the iron lung was invented in New York and used in Bellevue hospital by Philip Drinker and Louis Agassiz Shaw (Bellis, 2018). The iron lung enclosed the entire body, minus the head, and used a vacuum and pressure system to cause the lungs to inhale and exhale (Bellis, 2018). The first patients to use the iron lung were ones with chest paralysis due to the Polio virus (Bellis, 2018).

The aviation and science community also had an advancement with the emergence of the liquid fuel rocket in 1926 (Bellis, 2020). Robert Goddard's rocket helped to shape space travel (Bellis, 2019). He was the first to demonstrate that rockets can be propelled upward using liquid oxygen and nitrogen (Bellis, 2019).

There is not a household today that does not have a television, and society can thank Vladimir Zworykin (Bellis, 2019). His work with the iconoscope camera and the kinescope picture tube built the basis for John Baird, who developed the first live television in 1926 and later color transmission in 1928 (Bellis, 2019).

1931-1940

In the 1930s, a major technological innovation was launched with the invention of the electron microscope by Max Knoll and Ernst Ruska (Bellis, 2018). The importance of the electron microscope is that it gives scientists the ability to view objects the size of an atom (Bellis, 2018). The radio telescope was also invented in this decade by American Physicist Karl Jansky (Bellis, 2020). The radio telescope helped astronomers' further study and explore outer space (Hochheiser, 2020).

While the 20th Century had already seen many advancements in radio technology, the 1930s brought about the FM radio and tape recording (Bellis, 2020). The AM radio had already been in use but experienced a complication with static interference, resolved with Edwin Armstrong's invention of the FM radio (Bellis, 2019). The tape recorder was invented by Joseph Begun and changed how broadcasting and recording were developed (NIHF, 2022).

1941-1950

One of the world's greatest, and saddest inventions was the creation of the atomic bomb. Its creation started with the invention of the neutronic reactor by Enrico Fermi and later with the atomic bomb in 1945 (Bellis, 2020a). While it was a new and innovative invention for military warfare during World War II, the damage and destruction it caused led to significantly restricted manufacture and usage.

The medical community saw the creation of the kidney dialysis machine and the aqualung (diving regulator) in 1943 (Bellis, 2020). Willem Kolff helped pave the way to treating kidney disease resulting in saving countless lives (DPCEC, 2020). While not necessarily a medical device, the aqualung (i.e., SCUBA) allowed for deep underwater diving and exploration (NIHF, 2022). The microwave oven was also invented during this decade by Percy Spencer in 1945 (Bellis, 2020). Spencer used the invention of the radar from earlier in the century to create the microwave that is still used today.

1951-1960

The radio continued to flourish through this century with the invention of the transistor radio in 1953 by William Shockley, John Bardeen, and Walter Brattain (Trueman, 2015). The radio was different from others because it was portable (Trueman, 2015). The AT&T company invented the computer modem and integrated circuit, and in 1962 it became commercially available (Bellis, 2017). The integrated circuit aided in the creation of cellphones and computers that we use today. Jack Kilby was credited for its invention in 1958. (Aasen, 2018).

1957 was a big year for science with the launch of Sputnik, the first satellite into space by the Soviet Union (Trueman, 2015). According to NASA, its launch brought about many new developments in different spheres of life, such as the military, technology, and science. The following year brought the concept of the laser technology, although it was not built or tested until the end of the decade (Hogan & Hogan, 2019).

The medical field again saw some new technology in the endoscope and the pacemaker (Trueman, 2015). The fiberoptic endoscope was invented by Basil Hirschowitz in 1957. It helped improve the quality of endoscopic testing with its flexibility and lack of bulk which in turn lowered patient discomfort (Dittrick Museum of Medical History, 2022). The pacemaker was also invented during this time. Invented by Drs. Chardack and Greatbatch, who were Veteran Affairs physicians, the first implanted pacemaker helped the patient live another ten months. Later versions of their original design helped patients live 20 more years (The U.S. Department of Veteran Affairs, 2018).

1961-1970

Computer technology saw many new inventions during the 60s. Starting off the decade, the Netherlands brought to the world audio cassettes in 1962. Society was able to record and then go back and listen to previous recording (Harrison, 2019). Only three years later, society's relationship with music and audio recordings was changed again with the invention of the compact disc, also known as CDs, by James Russell (History Computer Staff, 2021). However, the compact disk did not gain popularity until the 1980s. The computer mouse was also invented during this time (Thrueman, 2015). Using radar technology, Douglas Engelbart developed and patented the first computer mouse, which created the path to other computer aids such as joysticks and trackballs (Hall, 2022). Another addition to the computer world was the invention of the Random Access Memory, commonly called RAM. This technology allowed data saved to be accessed randomly (History Computer Staff, 2021). Another piece of technology that was invented in the 60s that is still used today is the automatic teller machine (ATM), which allowed bank customers to withdraw cash from a machine (A&E Television Networks, 2009).

Major strides in medical technology occurred during this decade, with Dr. Denton A. Cooley conducting the first total artificial heart implant on April 4, 1969. The prosthetic heart was developed by Dr. Domingo Liotta that same year. The patient was able to live with the artificial heart for three days until a human heart was found. This jumpstarted the creation of mechanical devices to be used as placeholders until a viable organ can be found for transplant (Texas Heart Institute, 2021).

1971-1980

While the RAM chip was the first form of digital memory for computers, its downfall was that it was not portable. The floppy disc of 1971 solved that problem. This square plastic was widely used to save documents created on a computer and was convenient in its portability (PasteTech, 2016). The inkjet printer was also invented during this time. This printer was developed in 1977 and is still used today in households and industries (Pietrus, 2017).

In the medical field, 1977 made a huge leap in innovation with the invention of the MRI scanner. Invented by Raymond Damadian, the MRI scanner is still of medical benefit today in diagnosing and treating patients with diseases by producing excellent body images (Editorial Team, 2021).

It was in this decade that the personal computer and cell phone came to be. The All-in-one personal computer, or Apple I, was first created and sold in 1976 and made a start for other companies to develop and sell personal computers (PasteTech, 2016). 1973 saw the first- ever cell phone by Motorola engineer Martin Cooper. While the concept of a portable phone was already in existence with car phones, Cooper thought that the phone should be even more portable and easily accessible (PasteTech, 2021).

1981-1990

Further innovations in the medical field occurred in the 1980s. The scanning tunneling microscope, invented in 1981, provided a way to capture pictures at the size of an atom on a metal surface (Bellis, 2018). Scientists developed artificial skin and grafted it onto burn patients during this time as well. Loannia Yannas, a physical chemist, and John Burke, a physician, developed this artificial skin and were the first to test it on humans (NIHF, 2022).

The space shuttle and the World Wide Web were two other critical inventions of this decade. Even though the first noted space shuttle was invented in the 1970s, it was the shuttle Columbia in the 80s that was regularly launched into space. This led to the launch of the Hubble Space Telescope in 1990, which revolutionized the field of space exploration (A&E Television Networks, 2020). Of major prominence was the development of the World Wide Web, or Internet. Tim Berners-Lee, a British scientist, developed the Web, which was a method of organizing information and allowed for easy remote access from any computer (Trueman, 2015). This technology shifted and shaped the world and how we exist in the 21st Century.

1991-2000

While Berners-Lee created the World Wide Web, it was Google who revolutionized how we navigate the Internet. A major player in the search engine of today, Google changed how we learn, share ideas, and communicate (Bellis, 2020). 1992 brought about a new way of communicating through text messaging. Unlike phone calls, text messages allowed the users to send written messages via cellphones; its developer was Neil Papworth (Alexander, 2019).

The 1990s did not see many new and novel inventions but improved upon previous inventions. Instead of VHS tapes to watch movies, DVDs brought higher quality videos to consumers' homes. Society stopped using portable cassette tape players or boom boxes and started using small, compact, and digital mp3 players whose music was downloaded from the Internet. Instead of needing to go to arcades for games, children and adults alike were now able to play video games at home with early developers such as Nintendo and PlayStation (Ngyuen, 2019). Cellular/portable phones started to see an upgrade with palm pilots and Nokia phones, and accessibility worldwide also exploded. It is evident that the 20th Century had an abundance of novel inventions. Later iterations of many technologies made them better, faster, smaller, and crossed over into many industries.

The first two sections of this paper explain the magnitude of technological innovation that occurred in the 20th century. Not a year went by without someone either perfecting an existing invention or someone creating something brand-new. The first section merely listed all the technological inventions of the 20th Century, and the second section outlined how inventions impacted the medical field, and public health. The final section aims to provide a comparison of the two halves of the Century and to delve further into the impact that the inventions listed in section two had on the medical community.

1900-1949

The first half of the 20th Century saw a multitude and vast variety of inventions. Some were related to public health and the medical field and others whose purpose may not be specifically health-related but still had an impact on medical care.

In 1902 Willis Carrier invented the air conditioner (Bellis, 2021). His original calculations are still what is used today in our modern air conditioning systems (Bellis, 2021). His invention, especially with its ability to control temperature and humidity, helped to flourish many other industries (Bellis, 2021). Some limitations to his invention were that it worked in small areas, and the challenge was the ability to control the climate and humidity on a larger scale. His invention played a role in the medical field because many medications, equipment, and even lab samples need to be stored at specific temperatures and humidity levels.

1903 was an informative and innovative year in medical technology with the invention of the first electrocardiograph (National MagLab, 2014). It was invented by Willem Einthoven, a Dutch Physiologist (National MagLab,

2014). The waves patterns (P, Q, R, S, T), also developed by Einthoven, are still in use today (National MagLab, 2014). Einthoven's electrocardiograph was originally a 600-pound, five-person operated machine which posed challenges with moving. It was improved upon in later years through the 20th century to become the lightweight 12-lead tabletop or pocket size machine with even Bluetooth technology that is used today (National MagLab, 2014). Electrocardiographs are used by almost every medical professional to identify cardiac waves and their abnormalities, and it has saved countless lives by aiding in the early detection of heart attacks and arrythmias. As previously stated, its limitation in the early 1900s was that it was a large machine that required multiple personnel to operate.

1903 is when the Wright Brothers designed the airplane (Bellis, 2020). The design was rudimentary with a motor and an aircraft able to hold the motor's weight (Bellis, 2019). Several iterations later, the airplane has become a success for human engineering. The helicopter was also based on the airplane design. Paul Cornu, a French man is credited with this first design of a manned rotary wing aircraft. But it was Russian-American aviation engineer Igor Sikorsky's design that is most similar to the modern-day helicopter (Bellis, 2019). Airplane and helicopters are innovations that have significant impact on health, in particular public health. With airplanes and helicopters, patients in dire need can be quickly transported to healthcare facilities. It is not just the transportation of patients but also the transportation of medical supplies, medication, vaccines, and even organs for transplants, all of which are very time sensitive.

In 1904, John Ambrose Fleming invented the electron tube (Bellis, 2019). He was a successful inventor and is credited for starting the field of electronics. The purpose of these tubes was to regulate and control the movement of electrons and electricity. Now replaced by plasma and LCD technologies, the electron tube was a primary technology for radar and radio transmissions (Bellis, 2019). They were later improved upon in 1906 by Lee de Forest, who invented the triode, which aided in the magnification of electrons (Bellis, 2020). The limitations with the electron tubes, whether the diode or triode, was that they were bulky and fragile because they were made of glass. Nonetheless, the diode and triode were used in medical technologies for x-rays which help in identifying and diagnosing various diseases.

In 1913, Alfred W Wolf invented the first refrigerator for domestic use. While it was not as successful, many other versions were based upon Wolf's invention (Dittrick Museum of Medical History, 2022). Just like with air conditioning, the refrigerator is crucial in the medical field and public health in the form of storage. While the refrigerator is more commonly known for storing food in households, medical offices, clinics, and hospitals, they are also used to store medicines, vaccines, and laboratory samples. Major limitations of the refrigerator at the time of invention were that it was not electrical and used ammonia as a refrigerant.

The gas mask was invented in 1914 (Bellis, 2020). The benefit, especially during World War I (WWI) was that the wearer of the device would be protected from aerosolized toxins. The limitation at the time was that it was bulky and cumbersome to wear. Another limitation was that they did not fit properly on children. While they were initially invented to be used during WWI with the use of toxic gas, they benefited the medical field as they evolved into respirators and masks used today.

Naval engineer Earl C. Hanson used vacuum tube technology and designed the first hearing aids called Vactuphone in the early 1920s (Dahl, 2021). Hanson's breakthrough with hearing aids led to scientists and inventors perfecting the technology to what is sold today (Dahl, 2021). While Hanson's invention revolutionized care for the hearing impaired, a few limitations of his invention was that it was large, not portable, and required multiple batteries for use.

A breakthrough in medical technology was the iron lung. Invented in 1927 by Philip Drinker and Louis Agassiz Shaw (Bellis, 2018). The iron lung enclosed most of the person in a high-pressure vacuum that caused the lungs to inhale and exhale (Bellis, 2018). Poliomyelitis, which was prevalent at the time, saw a benefit in the use of the iron lung. While the iron lung allowed polio patients time to heal and the ability to maintain respiration, it did not allow for them to maintain their activities of daily living, and it caused some feelings of claustrophobia. The iron lung did, in fact, bring about the idea and technology for respirators and ventilators that we use today.

The electron microscope was invented in 1931 by Max Knoll, and Ernst Ruska (Bellis, 2018). The importance of the electron microscope is that it gave scientists the ability to view objects the size of an atom (Bellis, 2018). Main limitations of the electron microscope were that it did not allow the view of live specimens, and the images were in black and white. The impact that it had on public health is that it allowed for the visualization of tissue, cells, and organelles. The electron microscope was and still is greatly used when scientists and healthcare providers are investigating and visualizing small organisms such as viruses.

The tape recorder was invented by Joseph Begun and changed how broadcasting and recording were done (NIHF, 2022). It allowed for the ability to record notes or thoughts and then played them back. Its limitations were that the process with the original invention was complicated to record, but it revolutionized the dissemination of information. This greatly affected the medical and public health community because information from other scientists was readily available to the public.

Willem Kolff helped pave the way for treating kidney disease with the invention of the kidney dialysis machine (DPCEC, 2020). It helped to treat and extend the lives of those with acute kidney disease. The limitation of this invention was at the time it could only treat acute kidney disease, and not those with end-stage renal disease. Kolff's invention did help to start the development of the modern dialysis machine which treats all types of renal failure.

1950-2000

The medical field again saw new technology in the form of the endoscope and the pacemaker (Trueman, 2015). Basil Hirschowitz invented the fiberoptic endoscope in 1957 (Dittrick Museum of Medical History, 2022). It helped improve the quality of endoscopic procedures, decrease treatment time and patient discomfort, all while providing clearer images (Dittrick Museum of Medical History, 2022). A limitation of note was that the scope was fragile, and the fibers could potentially tear.

Another innovation in cardiac engineering was the creation of the pacemaker. Invented by Drs. Chardack and Greatbatch's, Veteran Affairs physicians in 1958, the first-generation implanted pacemaker improved the quality of life and extended the life of many patients. Subsequent designs had even more impressive gains on life expectancy and quality (The U.S. Department of Veteran Affairs, 2018). Pacemakers have extended life expectancy for patients with cardiac disease, but early versions of the device had inconsistencies and glitches.

In 1962 Netherlands inventors created the audio cassette. People were able to record, listen, and play back recording (Harrison, 2019). A few years later, the compact disc, also known as the CD, was created by James Russell (History Computer Staff, 2021). However, CD technology did not gain popularity until the 1980s. The cassette and CD made the distribution and accessibility of audio recordings much easier than before. This allowed medical and health professionals to record various aspects of healthcare for internal and societal distribution. This technology had its challenges, a main issue experienced was over time wear and tear of the CDs affected the sound quality.

Another addition to the computer world was the invention of the RAM, which allowed for data saved to be accessed randomly (History Computer Staff, 2021). RAM improved the functionality and quality of computers, which in turn improved data storage. The downside was that at the time, RAM was slow, unreliable, and costly.

Dr. Denton A. Cooley conducted the first total artificial heart implant on April 4, 1969, that was developed by Dr. Domingo Liotta. The patient was able to live with the artificial heart for three days until a human heart was found. This jumpstarted the creation of mechanical devices to be used as placeholders until a viable organ can be found for transplant (Texas Heart Institute, 2021).

In 1977 Raymond Damadian invented the MRI scanner, which is still of medical benefit today in diagnosing and treating patients with many diseases (Editorial Team, 2021). The MRI allowed for more reliable and accurate imaging and diagnosing diseases in a quick timeframe. However, the cost and access of the technology remains a challenge for many.

The All-in-one personal computer, or Apple I, was first created and sold commercially in 1976. This was the burgeoning of personal computers (PasteTech, 2016). 1973 saw the first cell phone by Motorola engineer Martin Cooper. Cooper's niche was to make the portable phone accessible from anywhere (PasteTech, 2021). These two inventions drastically accelerated the speed of technology and communication. Many healthcare devices used these technologies to strength healthcare services, knowledge, and access. Not without its challenges, the personal computer and cellphone are barriers to care and services for those without the savvy skill set and financial resources to purchase them.

Scientists in the 1980s significantly advanced burn care. Artificial skin and grafting were developed by Loannia Yannas, a physical chemist, and John Burke, a physician, they developed artificial skin and were the first to test it on patients (NIHF, 2022). The benefits of artificial skin were improved treatments and outcome for burn victims, shorten recovery times, and improved quality of life. Some of the challenges were the risk of infection as well as it required multiple surgeries.

Possibly the greatest technological achievement on the latter half of the Century, is the creation of the World Wide Web, or Internet. Tim Berners-Lee, a British scientist, developed the World Wide Web, which was a method of organizing information and allowed for easy remote access from any computer (Trueman, 2015). Google was also created some years later which help to navigate the Internet through technology called search engines (Bellis, 2020). Both the internet and google allow information to be accessible to everyone on a broad scale. Challenges with the Internet and search engines has led to information inaccuracies, reliability problems and disparities in access.

COMPARISON

It cannot be denied that the 20th century was a vastly innovative time in many different industries. The 20th Century began with the birth of airplanes, radio, cars, and the like. Those inventions both shocked and changed society for the better. By the end of the 20th Century, we saw vast improvements such as spaceships, computers, cellphones, and most of all the wireless Internet. Comparing the two halves of the Century, it can be summarized that the beginning of the century saw novel ideas and inventions while the latter half saw many improvements in these inventions. There was also a multitude of physical inventions in the first half of the Century, while the second was more virtual and microelectronic.

MOVING INTO THE 21ST CENTURY

The change of the millennium saw a large shift in society towards the internet and technology. In the past, the internet was not found in every household; now more that 90% of Americans have broadband internet (Hillyer, 2020) and most in the palm of their hands. This change was not just happening in the US but all around the world. Society has now shifted to where technology is vastly more portable and personal (Hillyer, 2020). The main takeaway moving from the 20th Century into the 21st is that everything is evolving and improving. In history, there has never been an invention that someone did not think that they could improve upon or even create something entirely new that could serve the same function. That is the mindset that the 21st Century has taken and even the next centuries after that.

REFERENCES

Aasen, E. (2018, November 28). *On this day in 1958, a TI engineer invented a chip that changed the world.* KERA News. Retrieved from https://www.keranews.org/health-science-tech/2014-09-12/on-this-day-in-1958-a-ti-engineer-invented-a-chip-that-changed-the-world.

Alexander, D. (2019, July 24). *17 of the most influential and interesting inventions from 1911-1920.* Interesting Engineering. Retrieved from https://interestingengineering.com/a-look-at-the-17-most-influential-inventions-from-1911-1920.

A&E Television Networks. (2009). *First ATM opens for business.* History.com. Retrieved from https://www.history.com/this-day-in-history/first-atm-opens-for-business#:~:text=On%20September%202%2C%201969%2C%20America's,to%20conduct%20basic%20financial%20transactions.

A&E Television Networks. (2020, February 15). *NASA unveils its first Space Shuttle, the Enterprise.* History.com. Retrieved from https://www.history.com/this-day-in-history/space-shuttle-unveiled.

Bellis, Mary. (2017, December 31). History of the Modem. Retrieved from https://www.thoughtco.com/history-of-the-modem-4077013.

Bellis, Mary. (2018, January 7). History of the Iron Lung - Respirator. Retrieved from https://www.thoughtco.com/history-of-the-iron-lung-respirator-1992009.

Bellis, Mary. (2018, February 28). Who Invented the Scanning Tunneling Microscope? Retrieved from https://www.thoughtco.com/scanning-tunneling-microscope-4075527.

Bellis, Mary. (2018, July 17). History of the Microscope. Retrieved from https://www.thoughtco.com/history-of-the-microscope-1992146.

Bellis, Mary. (2019, February 3). The History of Vacuum Tubes and Their Uses. Retrieved from https://www.thoughtco.com/the-history-of-vacuum-tubes-1992595.

Bellis, Mary. (2019, February 13). Invention and History of Rockets. Retrieved from https://www.thoughtco.com/invention-and-history-of-rockets-1992375.

Bellis, Mary. (2019, February 25). Biography of Charles Kettering, Inventor of the Electrical Ignition System. Retrieved from https://www.thoughtco.com/charles-kettering-electrical-ignition-system-4076281.

Bellis, Mary. (2019, May 07). Biography of Edwin Howard Armstrong, Inventor of FM Radio. Retrieved from https://www.thoughtco.com/edwin-howard-armstrong-1991244.

Bellis, Mary. (2019, May 10). The History of Radio Technology. Retrieved from https://www.thoughtco.com/invention-of-radio-1992382.

Bellis, Mary. (2019, July 3). Mechanical Television History and John Baird. Retrieved from https://www.thoughtco.com/television-history-john-baird-1991325.

Bellis, Mary. (2019, July 3). History of Flight: The Wright Brothers. Retrieved from https://www.thoughtco.com/history-of-flight-the-wright-brothers-1992681.

Bellis, Mary. (2019, October 4). History of the Helicopter. Retrieved from https://www.thoughtco.com/history-of-the-helicopter-1991899.

Bellis, Mary. (2019, December 3). History of the Escalator. Retrieved from https://www.thoughtco.com/history-of-escalator-4072151.

Bellis, Mary. (2020, January 24). 20th Century Invention Timeline 1900 to 1949. Retrieved from https://www.thoughtco.com/20th-century-timeline-1992486.

Bellis, Mary. (2020, January 29). The History of Google and How It Was Invented. Retrieved from https://www.thoughtco.com/who-invented-google-1991852.

Bellis, Mary. (2020, Mar 27). The History of Sonar. Retrieved from https://www.thoughtco.com/the-history-of-sonar-1992436.

Bellis, Mary. (2021, February 16). The Father of Cool - Willis Haviland Carrier and Air Conditioning. Retrieved from https://www.thoughtco.com/willis-haviland-carrier-air-conditioning-4078668.

Dahl, N. (2021, March 4). *Hearing aid history: Ear trumpets, European Royalty, & earbuds*. Hearing Health Foundation. Retrieved from https://hearinghealthfoundation.org/blogs/hearing-aid-history-ear-trumpets-european-royalty-earbuds.

Dittrick Museum of Medical History. (2022). *Hirschowitz fiberoptic endoscope, 1960*. Retrieved from https://artsci.case.edu/dittrick/online-exhibits/explore-the-artifacts/hirschowitz-fiberoptic-endoscope-1960/.

DPCEC. (2020, April 28). *A brief history of dialysis*. Dialysis Patient Citizens Education Center. Retrieved from https://www.dpcedcenter.org/news-events/news/a-brief-history-of-dialysis/.

Editorial Team. (2021, March 16). *The invention of Magnetic Resonance Imaging (MRI)*. IEC. Retrieved from https://www.iec.ch/blog/invention-magnetic-resonance-imaging-mri#:~:text=Raymond%20Damadian%2C%20the%20inventor%20of,a%20human%20being%20in%201977.

Electrocardiograph – 1903. National MagLab. (2014, September 20). Retrieved from https://nationalmaglab.org/education/magnet-academy/history-of-electricity-magnetism/museum/electrocardiograph.

Hall, M. (2022). *Douglas Engelbart*. Encyclopedia Britannica. Retrieved from https://www.britannica.com/biography/Douglas-Engelbart.

Harrison, B. (2019, July 6). *Design moment: Cassette Tape, 1962*. The Irish Times. Retrieved from https://www.irishtimes.com/life-and-style/homes-and-property/interiors/design-moment-cassette-tape-1962-1.3945229.

History.com. (2010, March 4). *First liquid-fueled rocket*. History.com. Retrieved from https://www.history.com/this-day-in-history/first-liquid-fueled-rocket.

History Computer Staff. (2021, December 23). *Compact disc explained - everything you need to know*. History Computer. Retrieved from https://history-computer.com/compact-disc/.

History Computer Staff. (2021, January 4). *Dram - history of Dynamic Random Access Memory*. History Computer. Retrieved from https://history-computer.com/dram-history-of-dynamic-random-access-memory/.

Hochheiser, J. (2020, September 24). *The history of the Radio Telescope*. Mini-Circuits. Retrieved from https://blog.minicircuits.com/the-history-of-the-radio-telescope/#:~:text=The%20radio%20telescope%20has%20come,be%20used%20for%20space%20research.

Hogan, M. R., & Hogan, H. (2019, June 6). *A history of the Laser: 1960 - 2019*. Photonics Media. Retrieved from https://www.photonics.com/Articles/A_History_of_the_Laser_1960_-_2019/a42279.

Lessing, L. P. (2021, December 14). Edwin H. Armstrong. Encyclopedia Britannica. https://www.britannica.com/biography/Edwin-H-Armstrong

Moore, L. (2017, January 27). *Sir James Mackenzie - 1853-1925*. Hektoen International. Retrieved from https://hekint.org/2017/01/27/sir-james-mackenzie-1853-1925/.

National Academy of Science (2022, September 14). *Air Conditioning and Refrigeration History - part 3*. Retrieved from http://www.greatachievements.org/?id=3862.

NAHF. (2020). *Fokker, Anthony*. Fokker, Anthony: National Aviation Hall of Fame. Retrieved from https://www.nationalaviation.org/our-enshrinees/fokker-anthony/#:~:text=During%20World%20War%20I%2C%20Fokker%20perfected%20a%20synchronized%20machine%20gun,route%20to%20England%20in%201920.

Nguyen, Tuan C. (2019, March 3). The Top Inventions of the 1990s. Retrieved from https://www.thoughtco.com/1990s-inventions-4147456.

NIHF. (2022). *NIHF inductee Charles Seeberger, who invented the escalator*. National Inventors Hall of Fame. Retrieved from https://www.invent.org/inductees/charles-d-seeberger.

NIHF. (2022). *NIHF inductee Ioannis v. Yannas Invented Artificial Skin.* NIHF Inductee Ioannis V. Yannas Invented Artificial Skin. Retrieved from https://www.invent.org/inductees/ioannis-v-yannas.

PasteTech. (2016, July 6). *7 tech advancements from the 70s that changed the world.* pastemagazine.com. Retrieved from https://www.pastemagazine.com/tech/7-tech-advancements-from-the-70s-that-changed-the/.

Pietrus, B. (2017, September 21). *The history of Printing.* 1ink.com. https://www.1ink.com/blog/the-history-of-printing/#:~:text=The%20inkjet%20printer%20was%20invented,like%20a%20dot%20matrix%20printer.

Texas Heart Institute. (2021, April 16). *50th anniversary of the world's first total artificial heart.* https://www.texasheart.org/50th-anniversary-of-the-worlds-first-total-artificial-heart/

The U.S. Department of Veteran Affairs. (2018, August 2). The invention of the cardiac pacemaker. https://www.research.va.gov/research_in_action/The-invention-of-the-cardiac-pacemaker.cfm

Trueman, C. N. (2015, April 21). *Radar.* History Learning Site. https://www.historylearningsite.co.uk/world-war-two/world-war-two-in-western-europe/battle-of-britain/radar/

Chapter 4

Industrial Revolutions, Industrialization, and Impacts on Human Environment

The Imperativeness of Environmental Education

By Tolu O. Isele, Bimpe E. Alabi & Kofo A. Aderogba

INTRODUCTION

God spent the first five days of the creation week preparing the environment for human beings who were to follow. He placed Adam and Eve in a garden (Gen. 2:8, 15), the garden of Eden, and gave them the oversight of all conceivable creatures – plants, animals, including all physical material nonliving things (Zondervan, 2011). These circumstances offered them the opportunity for learning and enjoyment and the chance to develop as God's representatives on the planet Earth. Incidentally, sin changed God's preferred plan (Ijezie, 2021). Ijezie (2021) opined that along the lines of the vicissitudes of the human vocation within the garden were: structures in a schema of crisis-resolution (Genesis 2:4b-5); crisis situation - the earth without vegetation and human labor (Genesis 2:6-25); the planting of a garden and creation of the Adam for labor (Genesis 3:1-21); complication of the situation: human abuse of the garden (Genesis 3:22-24); and final resolution: expulsion from labor in the garden to labor on the dry ground. However, the more the number of individuals, the more the needs; and the more the development of the sophistication of the processes of development and technology to increase production (and services). Thus, over the centuries came the stage of hunting and gathering, farming and cultivation of crops, and invariably Industrial Revolution, and gradually the contemporary industrial development.

The environment has become hostile. Hard drudgery replaced pleasurable work. Selfishness ruled. Exploitation of the resources of the planet Earth began and continues to this day. Forests have been cut down. Water courses have been changed; and pollutions of water, soil, air, noise, and others are widespread. The wealth of the water, soil, air, and others have been squandered thoughtlessly. Sunlight, though crucial to wellbeing, it can as well become an adversary to health and wellbeing on excessive exposure of it to the planet Earth and the creations on it (Ijezie, 2021). But through all these, the plant Earth still supports human existence despite the endless abuses.

The thrust of this chapter is to address the industrial revolution, industrialization, and impacts on natural environment of man. It outlines the significance of aspects of environmental education for sustainable environment. It is orchestrated into eight parts: This introduction; definition of concepts; industrial revolution and industrialization; consequences of the industrial revolution and industrialization; pollutions due to industrial waste; education for sustainable development; education for sustainable environment; and conclusion.

CONCEPTUAL UNDERPINNING

For clarity of purpose, some salient concepts have been highlighted: the concept of the natural environment, human-environment interaction, and sustainability. These have been given due recognition as relevant to the theme of this chapter.

The Concept of Natural Environment

The natural environment is the natural world. It includes all living and non-living things occurring naturally. The term natural environment is most frequently applied to the Earth or some parts of the Earth. It incorporates the interaction of all living species (flora and fauna), climate, weather, and natural resources that affect human survival and economic activities (Johnson et al., 1997). The concept can be distinguished into two components: (1) Comprehensive ecological components that function as natural systems without human interference, including all flora, microbes, earth (soil), rocks, atmosphere, and natural phenomena that occur within their boundaries and their nature; and (2) Worldwide natural resources and physical phenomena that lack precise borders, such as air, water, and climate, in addition to energy, radiation, electric charge, and magnetism, that did not originate from cultured human activities.

Contrasting to the natural environment is the built, man-made environment. These are where humans have essentially altered settings, such as settlement settings and agricultural land adaptation. The natural environment is greatly changed into a basic human environment. Incidentally, through those acts which seem less life-threatening, such as building a hut or laying of water pipes in the forest, the altered environment becomes non-natural and artificial. Nevertheless, many animal-built things to provide a better environment for themselves; however, they are not human, hence beaver dams, and the works of mound-building termites, are thought of as natural (Aderogba, 2016; Aderogba, 2021; Aderogba & Bankole, 2016; Aderogba et al., 2021; Awobajo, 1980; Goudie, 2000).

The surface of the planet Earth contains a number of diverse natural environments that include the flora, fauna, rocks, minerals, and atmosphere that make up a single ecological system spanning a large area. The natural environments often overlap and compete with each other and human-built environments, such as cities, manufacturing plants, and parks.

Thus, different types of natural environment include oceans, grasslands, tundra, rainforests, and deserts, and others. Each environment has unique soils, climates, water systems, and weather phenomena capable of supporting the life forms in that environment. They are constantly changing and can be greatly impacted by physical and chemical changes on the planet Earth. The diversity of natural environmental types reflects the rich variety of flora and fauna species on the planet Earth and the different conditions necessary for them to thrive.

The term ecosystem, meaning a system of living organisms in an area that depends on the proximate non-living resources to survive and reproduce, is often used interchangeably with the term natural environment. While this is correct in some cases, it is important to note that a single natural environment can also contain multiple ecosystems. The oceans are a good example of this: the flora and fauna that thrive in coastal regions require different living conditions than the organisms found in the deep ocean (Ask Media Group, 2021).

Rain forest environments are one type of natural ecosystem. In total, rain forests produce 40 percent of the oxygen on the planet Earth, despite only making up 6% of the surface of the Earth. They are composed of four layers: emergent, upper canopy, understory, and the forest floor. While these layers are part of one environment, the flora and fauna differ from layer to layer (Aderogba, 2021; Aderogba et al., 2021; Reference.com, 2021). Rain forests get their name from the fact that they are almost completely self-watering. Each tree that reaches the canopy level is able to release about 200 gallons of water into the air per year. This circumstance creates an enduring cloud that hangs low over the canopy, helping the forest stay hydrated between periods of rainfall (Reference.com, 2021). This creates a fertile environment for plants, with two-thirds of the total plant species on Earth growing in the rainforest. In recent years, rainforest environments have been threatened by deforestation, and scientists recommended that further logging in rain forests could have long-term negative effects on the climate of the planet Earth (Aderogba, 2021; Reference.com, 2021).

Ocean environments are found all over and are the largest natural environment. More than 70% of the surface of the Earth is covered by oceans, and they contain a total of 97% of our water supply. The world's oceans support many unique habitats - phytoplankton, kelp, and seaweed thrive at the surface, and tube worms, mussels, and clams live on the deep-sea floor. This is possible due to hydrothermal vents that pump minerals from the Earth's crust into the ocean.

Desert environments can be found on every continent and can be either very hot and sandy or very cold and frosty. While desert environments have a reputation as inhospitable places to live, about one-sixth of human population lives in a desert environment. Deserts are distinguished by the fact that they are likely to regularly lose more moisture through evaporation than they receive through precipitation. Flora and fauna in the desert environments have found ways to adapt to the harsh living conditions. Plants that may not receive water for several years at a time have adapted by either finding water with their roots deep underground or by being able to store reserves of water for long periods of time. In hot desert environments, many animals avoid the heat by living a nocturnal lifestyle, only searching for food and water at night.

Grasslands are environments formed when an area receives too much precipitation to be classified as a desert but not enough precipitation to support a forest environment. They are characterized by their most populous form of plant - grasses. The two types of grasslands are tropical and temperate, with tropical grasslands usually falling in the southern hemisphere. Tropical grasslands have both a dry season and a rainy season. Because these tropical environments receive more rain than temperate grasslands, their grasses can grow up to 2.13 meters tall. Temperate grasslands are characterized by shorter grasses and have both a growing and dormant seasons. During the dormant season, no grass grows due to cold temperatures.

Tundra environments can be found around the tops of mountains and in the Arctic. Mountain Tundra environments are home to some types of sheep, birds, and mountain goats, while the Arctic tundra houses animals like polar bears, caribou, and Arctic foxes. The tundra is one of the most inhospitable environments for plants and animals, with a cold climate, low rainfall, frequent winds, long winters, and short summers. The tundra environment is highly vulnerable to the effects of global warming, as species are being displaced from disappearing Arctic permafrost into tundra environments, disrupting the ecological balance of the tundra.

But absolute natural environments on Earth seem to be no more, and spontaneity typically contrasts in a gamut, from absolute (100.00%) natural in one extreme to zero (0.00%) natural in the other. The enormous environmental deviations of humankind in the Anthropocene have primarily produced all non-natural environments: including from climate change, biodiversity loss, and pollution from plastic and other chemicals in the air and water. Indeed, different aspects or components of an environment can be considered, but their degree of naturalness is not uniform (Symons, 1979). For instance, in an agricultural field, the mineral configuration and the structure of its soil are similar to those of uninterrupted forest soil, but the structure is quite different.

The natural environment is frequently used as a substitute for habitat, such as when we say that the natural environment of three-toed sloth is the high rain forest. However, human-environment interaction has been humongous in so many instances (Aderogba, 2016; Aderogba & Bankole, 2016; Ask Media Group, 2021; Refrence.com, 2021).

The Concept of Human-Environment Interaction

Human-environment interaction is the interrelationship between humans and the ecosystem of the planet Earth. Also, it is related to the ways people adapt and modifies nature, often to suit their purposes (Macmillan, 2020; McCloy, 2019). In other words, Human-environment interaction relates to a phenomenon that explains how humans interact with the environment to fulfill their needs and how the environment responds to these interactions. Moreover, it also describes how humans have adapted to their environment for survival.

Since geography shares a close relationship with humans and the environment, it is essential to discuss it for a better understanding of the human-environment interactions. There are five themes of geography, which include location, place, movement, human-environment interaction, and region: Location - relates to the geographical position of a specific place on the Earth. Latitude and longitude are the crucial parameters to locating a place; Place - discloses the physical specifications of a region such as wet, dry, hot, or cold; Movement - relates to the transportation and immigration patterns in a particular region; Region - specifies an area showing identical characteristics; and an all-inclusive discussion regarding the concept of human-environment interaction and its consequences as discussed below. However, there are three basic types of this interrelation: (1) The way human modifies the environment. This type includes not only a positive but also a negative impact on the planet Earth. For example, it concerns diking and borehole drilling; (2) Human dependence on the environment - People cannot live without the usage of such important natural resources - food, water, coal, wood, oil, and others; and (3) The way people use the environment to satisfy their own needs. Human civilization constantly tries to adapt nature to obtain better and more comfortable living conditions.

Thus, nature is essential for the well-being of human civilization. It goes with the saying that we affect the environment by using natural resources and modifying the ecosystems (Aderogba, 2016; Aderogba et al., 2021; Awobajo, 1980; MacMillan, 2020). Moreover, such activities may damage nature and people living in some areas. Despite these, there are a number of facts about human-environment interaction proving that people can get many of the benefits without causing damage to the ecosystem (Aderogba et al., 2021; MacMillan, 2020). The followings are interesting examples:

- Waste Dumping and Litter Pollution - Probably, this example has the most significant number of adverse effects. Litter causes ecological problems such as river pollution and the spread of various diseases dangerous to animals and humans (and even plants). Unfortunately, the recycling technologies are not perfect, and more than 11 tons of recyclable garbage make their way into landfills annually.
- Usage of Natural Resources - People use tons of metals, gas, and timber every day. At the same time the human population keeps on growing and that is why the demand for natural resources is increasing rapidly. Again, the socio-economic advancement leads to the rise of utilization of natural resources. Due to all these factors, human is faced with such serious ecological problem as the exhaustion of environmental assets.
- Deforestation - Every year, people cut down forests and do not plant new trees. Such human interaction with nature leads to complete destruction of the environment. Moreover, deforestation results in the death of wild animals; and it increases the level of greenhouse gases, soil erosion, and floods. Trees helping in regulating the water cycle are indirectly destroyed; and without them, less water will be available in the *Water Cycle*. Invariably, it becomes impossible to grow crops due to the dry ground that will result unless complemented and or supplemented.
- Graffiti and Usage of Aerosols - Aerosols are poisonous to nature. They have an impact on climate change and promote global warming. Even CFS-free sprays should be understood as substances that can boost the warming effect of the greenhouse gases in the atmosphere. In addition, the chemicals thrown away by aerosols are not just destroying the ecosystem of the planet Earth, but as well causing such serious illnesses as asthma.
- Borehole Drilling - It goes without saying that sometimes the illegal oil drilling may generate the substances that get into the ocean and poison a number of aquatic plants and animals. Also, the natural gas output is often the reason that leads to the complete destruction of various ecosystems. In

addition, drilling generates massive methane that affects the environment by contributing to global warming.

But the Human-Environment Interaction has some advantages, directly and indirectly, by constantly trying to help Mother Nature. These showcase examples of positive human-environment interaction. There are chances to reduce environmental impact and protect nature from ecological catastrophes. The followings are good examples:

- Conservation of Water Resources - Water is vitally important for the whole human civilization, and we cannot live without it. That is why more communities develop and implement such techniques as harvesting of rainwater, management of the watershed, and drip irrigation. Besides, they suppress the throwing of industrial and domestic wastes into the rivers and oceans as well as start various clean-up projects. With such an attitude, water pollution may be prevented even in industrial countries like China, Britain, and others.
- Recycling - It is an integral part of positive human-environment interaction. Interestingly, nature recycles everything, including dead animals or fallen trees by returning them to the soil to grow up more of the trees and other plants. That is why humans interacting with the environment should also follow this example and remake old products into new ones. Parenthetically, the recycling industry needs a lot of improvements. The United States recycles only 35% of the municipal waste which is approximately 91 million tons. Whereas the country produces more than 250 million tons annually. But the recycling industry continues to develop.
- Development of Sustainable Infrastructure - The main task of sustainable infrastructure is to ensure that the building of new roads, bridges, or other projects will cause the minimum damage to the environment. In addition, it includes the usage of eco-friendly and durable construction materials as well as the development of smart building technologies. A well-known example of sustainable urban infrastructure is Masdar City, situated in the United Arab Emirates (UAE).
- Ecotourism – Because humans depend on the environment, travel agencies produced the idea to mix tourism and the desire to protect the environment. Thus, building ecological awareness and a respectful attitude to the local culture. This is not just a great way to reduce the environmental damage caused by usual tourists but also to support the people living in some communities. For instance, the government of Madagascar managed to reduce its 81% poverty rate by encouraging ecotourism.
- Creation of National Parks – Just like ecotourism, the National Parks and special zones provide various animals and plants with protection from

human influence. But such places turn into tourist attractions where people can enjoy nature without damaging it, and they are laboratories for research and development.

However, the question could be asked: How does Human-Environment interaction specifically affects lives? As a matter of fact, as earlier noted, man depends on nature as the nature depends on man. Where people do not take care of the local environment, man may suffer from different diseases. For example, health may be damaged by polluted air or water. According to the statistics, 9 out of 10 people worldwide breathe impure air that causes such serious illnesses as lung cancer and asthma (Aderogba, 2016; MacMillan, 2020).

Following the eco-friendly tips and avoiding contributing to the environmental impact, one will see that life is changing for the better. It should be remembered that diet, clothing, and even general condition depend on the environment one lives (MacMillan, 2020). According to her, recent research has shown that workers in green-certified offices have 6.4% higher sleep quality scores than those that work in a completely different environment.

There are many negative ways humans interact with the environment, and each of them may affect lives in the long run. But preventing all negative implications means man should always respect nature and minimize the consumption of unrenewable resources.

In other words, nature has provided mankind with an environment full of resources that can simplify lives. Humans have destroyed the perfect balance and damaged the ecosystem, but the good news is that there is always a chance to prevent ecological disasters and maintain only positive human-environment interactions. Man can modify the planet Earth where he lives. But, without forgetting about the consequences of such changes. Man is responsible for nature and the future of the Earth. If he does not want the civilization to be faced with the destruction, he should do the best to live coordinated with nature, therefore. Although human-environment interactions benefit humans, it is impossible to ignore their far-reaching impacts on the environment. As earlier noted, one of the several consequences is climate change, which serves as a serious threat to humans as well as the generality of the environment. Therefore, the temperature of the earth is increasing at a fast pace. There is a potential threat that various coastal areas would submerge due to rising sea levels. Unrestrained use of fossil fuels and cutting down forests have caused irreparable damage to the atmosphere. Hence, humans are responsible for the current devastation of the environment facing the challenge of climate change which puts the future of the next generations at risk (McCloy, 2019).

The Concept of Sustainability

Sustainability is an extensive policy concept in the international public treatise, and it is often perceived in terms of three "dimensions" or "pillars," that is environmental, economic, and social (Purvis et al., 2019); see Figure 4.1. The original semantic meaning of "sustainability" refers to the ability to continue over a long period of time. A closely related and overlapping concept is that of "sustainable development." The United Nations Educational, Scientific and Cultural Organization (UNESCO) formulated a distinction that: "Sustainability is often thought of as a long-term goal (i.e., a more sustainable world), while sustainable development refers to the many processes and pathways to achieve it" (UNESCO, 2015). According to the United Nations General Assembly (1987) in the "Brundtland Report" *Our Common Future*, sustainable development is development that "meets the needs of the present without compromising the ability of future generations to meet their own needs."

The terms "sustainability" and "sustainable development" are increasingly becoming popular, and the issues often considered are complex (Raskin, 2002). The discussion of the environmental dimension of sustainability is repeatedly centered around prevailing issues. For example, the dominant issues since the early 2000s are climate change, biodiversity loss, environmental degradation, the biogeochemical fluxes of nitrogen and phosphorus, and pollution, especially plastic pollution, and others. The concept has also

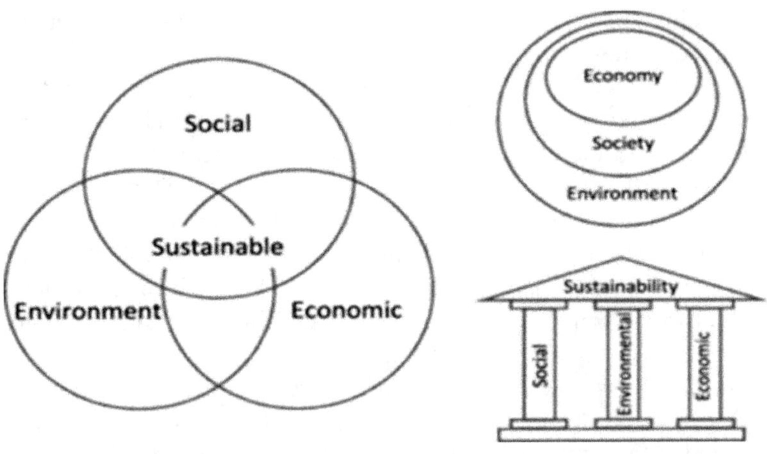

Figure 4.1 Commonly used schematics of the tripartite description of sustainability: Left, typical representation of sustainability as three intersecting circles. Right, alternative depictions: literal pillars and concentric circles approach. Purvis, B., Mao, Y. & Robinson, D. (2019). Three pillars of sustainability: In search of conceptual origins. Sustainability Science, 14(3), 681-695.

been described as an "exhausted roadmap" because our consumer societies are socially and ecologically self-destructive (Aderogba, 2013; Blühdorn, 2017). Stirring towards sustainability can involve social challenges such as individual lifestyles and ethical consumerism. Sustainable living approaches can reduce environmental impacts by altering the built environment to create more sustainable cities. That would include sustainable transport, zero-emission housing, sustainable architecture, and circular flow land use management.

Literature has many other explanations/definitions in literature (Aderogba, 2013; Blühdorn, 2017; University of California, 2021). The University of California (2021) brings together a community of students, faculty, and staff under the concept. They aver that "sustainability is the balance between the environment, equity, and economy." It is the practice of using natural resources responsibly so that they can support both present and future generations.

Forests are one natural resource that sustainability groups are focused on conserving. Forests made up about 30% of the Earth's land mass in 2015, but that number is at risk of decreasing. In 2018, studies showed that approximately 18 million acres of trees are cleared each year for lumber or agriculture (University of California, 2021). Deforestation destroys the habitat of other important organisms, including fruits, animals, and mushrooms that humans use for food or medicine (Aderogba, 2016, 2021). Deforested land also increases soil erosion, limiting the productivity of tree growth and others. The goal of sustainable forestry is to preserve forest ecosystems. Sustainable practices include planting new seedlings in deforested areas and reducing the number of trees cut down each year.

Specifically, as the human population grows, we must reduce the depletion of forests, precious metals, and other natural resources. The world population is expected to increase from 7.6 to 9.8 billion people by 2050 (United Nations, 2017). There has to be plan for the increase. Many scientists are already researching renewable energy sources to decrease our dependency on fossil fuels: The United States military is evaluating a vehicle fuel made from algae rather than petroleum: The United States military is testing a vehicle fuel made from algae rather than petroleum. Furthermore, concerns over freshwater sustainability have led to the development of rainwater harvesters, desalination machines, and even more incredible inventions. Some companies have produced drinking straws that filter bacteria, for example, SimPure and LifeStraw, so people can safely drink water from contaminated ponds and lakes.

But what are the features of the industrial revolution and industrialization and man- environment interaction; and what have been the sustainability drives of communities to "meet the needs of the present without compromising the ability of future generations to meet their own needs." The followings expatiate.

INDUSTRIAL REVOLUTIONS AND INDUSTRIALIZATION

Industrial Revolutions

The earliest recorded use of the term "Industrial Revolution" appears to have been in a letter of 6 July 1799 written by a French envoy, Louis-Guillaume Otto, announcing that France had entered the race to industrialize (Wrigley, 2018). In his 1976 book Keywords: *A Vocabulary of Culture and Society,* Raymond Williams states in the entry for "Industry": "The idea of a new social order based on major industrial change was clear in Southey and Owen, between 1811 and 1818, and was implicit as early as Blake in the early 1790s and Wordsworth at the turn of the [19th] century." The term Industrial Revolution applied to technological change that was becoming more common by the late 1830s, as in the description of Jérôme-Adolphe-Blanqui's (1837) la révolution industrielle.

Friedrich Engels, quoted by Hudson (1992) in *The Condition of the Working Class in England in 1844,* spoke of "an industrial revolution, a revolution which at the same time changed the whole of civil society." Though Engels authored his book in the 1840s, it was not translated into English until the late 1800s, and his expression did not enter everyday language until then. Credit for popularizing the term may be given to Arnold Toynbee, whose 1881 lectures gave a detailed account of the term (Hudson, 1992).

Economic historians and authors such as Mendels and Kridte argue that the proto-industrialization in parts of Europe, Islamic world, Mughal India, and China created the social and economic conditions that led to the Industrial Revolution, thus causing the Great Divergence (Broadberry et al., 2017; Broadberry & Gupta, 2005; Ogilvie, 2008). Some historians have argued that the economic and social changes occurred gradually and that the term "revolution" is a misnomer. This assertion is still a subject of debate among some historians (Craft, 2005, 2011).

However, the Industrial Revolution was the transition to new manufacturing processes in Great Britain, continental Europe, and the United States in the period from about 1760 to sometime between 1820 and 1840 (Wrigley, 2018). This transition included going from hand production methods to machines, new chemical manufacturing, and iron production processes, increasing use of steam power and waterpower, development of machine tools and the rise of the mechanized factory system. The revolution also led to an extraordinary rise in the rate of population growth.

It began in Great Britain, and many of the technological and architectural innovations were of British origin (Wrigley, 2018). By the mid-18th Century, Britain was the leading commercial nation in the world (Reisman, 1998), controlling a global trading empire with colonies in North America and the Caribbean, and with major military and political hegemony on the Indian sub-continent, particularly with the proto-industrialized Mughal Bengal, through the activities of the East India Company (Esposito, 2004; Landes, 1999; Ray, 2011; Tong, 2016). The development of trade and the rise of business were among the major causes of the Industrial Revolution (Landes, 1969; 1999).

The revolution marks a significant turning point in history; almost every aspect of daily life was affected. In particular, average income and population began to exhibit unprecedented, sustained growth. Some economists have said the most important effect was that the standard of living for the general population in the western world began to increase consistently for the first time in history, although others have said that it did not begin to meaningfully improve until the late 19th and 20th centuries (Feinstein, 1998; Lucas, 2002; Szreter & Mooney, 1998).

Gross Domestic Product (GDP) per capita was broadly stable before the Industrial Revolution and the emergence of the modern capitalist economy. At the same time, the Industrial Revolution began an era of per-capita economic growth in capitalist economies (Lucas, 2002). Economic historians agree that the onset of the Industrial Revolution is the most important event in the history of humanity since the domestication of animals and plants (McCloskey, 2004).

The precise start and end of the revolution are still debated among historians, as it is the pace of economic and social changes (Berg & Hudson, 1999). But rapid industrialization began in Britain, starting with mechanized spinning in the 1780s, with high rates of growth in steam power and iron production occurring after 1800 (Broadberry, & Gupta, 2005). Mechanized textile production spread from Great Britain to continental Europe and the United States in the early 19th Century, with important centers of textiles, iron, and coal emerging in Belgium and the United States and later textiles in France (Landes, 1969).

An economic recession occurred from the late 1830s to the early 1840s when the adoption of the early innovations of the Industrial Revolution, such as mechanized spinning and weaving, slowed and their markets matured. Innovations developed late in the period, such as the increasing adoption of locomotives, steamboats and steamships, hot blast iron smelting, and new technologies, such as the electrical telegraph, widely introduced in the 1840s and 1850s, were not powerful enough to drive high rates of growth. Rapid economic growth began to occur after 1870, springing from a new group of innovations in what has been called the Second Industrial Revolution. These innovations included new steel-making processes, mass-production, assembly lines, electrical grid systems, large-scale manufacturing of machine tools, and increasingly advanced machinery in steam-powered factories (Hunter, 1985; Landes, 1969). However, six factors facilitated industrialization:

1. High levels of agricultural productivity to provide excess manpower and food.
2. A pool of managerial and entrepreneurial skills.
3. Available ports, rivers, canals, and roads to cheaply move raw materials and outputs.
4. Natural resources such as coal, iron, and waterfalls.
5. Political stability and a legal system that supported business; and
6. Availability of financial capital to invest.

As soon as industrialization began in Great Britain, new factors started springing-up: the eagerness of British entrepreneurs to export industrial expertise and the willingness to import the process. Britain met the criteria and industrialization starting in the 18th century; and exported the process to Western Europe (particularly Belgium, France, and the German States) in the early 19th century. The United States copied the British model in the early 19th Century, and Japan copied the Western European models in the late 19th century (Deane, 1973; Rider, 2007).

The commencement of the Industrial Revolution is closely linked to a small number of innovations (Bond et al., 2003), beginning in the second half of the 18th century. By the 1830s, the following gains had been made in important technologies: textiles, steam power, iron making, the invention of machine tools, large-scale production of chemicals (Sulfur Acid, alkali, calcium hypochlorite (bleaching powder), cement, gas lighting, paper machine, agriculture, mining, transportation (canal and improved waterways, roads, railways), and others. Suffice to say that the revolution led to some social effects that were remarkable: factory system, the standard of living, food and nutrition, housing and sanitation, water supply, literacy (and industrialization), clothing and consumer goods, population increase and urbanization, women and family lives, labor conditions, factories and urbanization, child labor, organization of labor, capitalize, the shift in production centers and gravity, share in the total world manufacturing output, and others.

Beyond Great Britain, the Industrial Revolution in Continental Europe came later than in Great Britain. It started in Belgium and France, then spread to the German States by the middle of the 19th century. In many industries, this involved the application of technology developed in Britain in new places. Typically, the technology was purchased from Britain, or British engineers, and entrepreneurs moved abroad in search of new opportunities (Milward

& Berrick, 1977; Milward & Saul, 1973). Sooner than expected, Continental Europe, Australia-Hungary, Belgium (the second country in which the Industrial Revolution took place and the first in Continental Europe), France (where the revolution did not correspond to the main model followed by other countries), Germany (that became dominant in the chemical industry of the world in the late 19th Century), Sweden (that experienced two parallel economic movement – an agricultural revolution with large agricultural estates, new crops, and farming tools and commercialization of farming and a proto-industrialization with small industries being established in the country side), Japan, United States, and others came on board.

Steel is often cited as the first of several new areas for industrial mass-production, which are said to characterize a "Second Industrial Revolution." beginning around 1850. However, a method for mass manufacturing of steel was not invented until the 1860s. However, the New Industrialist movement advocates for increasing domestic manufacturing while reducing on a financial-based economy that relies on real estate and trading speculative assets. New Industrialism has been described as "supply-side progressivism" or embracing the idea of "Building More Stuff" (Klein, 2021). New Industrialism developed after the China Shock that resulted in lost manufacturing jobs in the United States (US) after China joined the World Trade Organization in 2001. The movement gained strength after the reduction of manufacturing jobs during the Great Recession and when the United States was not able to manufacture enough test kits or facemasks during the COVID-19 pandemic (Smith, 2022). New Industrialism calls for building enough housing to satisfy demand in order to reduce the profit in land speculation, invest in infrastructure, and develop advanced technology to manufacture green energy for the world. The belief is that the United States is not building enough productive capital, and it should invest more in economic growth (Smith, 2022).

Industrialization

As earlier noted, textiles were the dominant industry of the Industrial Revolution in terms of employment, the value of output, and capital invested. The textile industry was also the first to use modern production methods (Landes, 1969). After the last stage of the Proto-industrialization, the first transformation from an agricultural to an industrial economy is known as the Industrial Revolution and took place from the mid-18th to early 19th Century in certain areas in Europe and North America; starting in Great Britain, followed by Belgium, Switzerland, Germany, and France (Griffin, 2010). Characteristics of the early industrialization were technological progress, a shift from rural work to industrial labor, financial investments in new industrial structures, and early developments in class consciousness and theories related to this

(Sampath, 2016). Later analysts have called this the First Industrial Revolution (Pollard, 1981).

The "Second Industrial Revolution" labels the later changes that came about in the mid-19th Century after the refinement of the steam engine, the invention of the internal combustion engine, the harnessing of electricity, and the construction of canals, railways, and electric-power lines. The invention of the assembly line gave this phase a boost. Coal mines, steelworks, and textile factories replaced homes as the place of work (Buchheim, 1994; Jones, 2003).

By the end of the 20th Century, East Asia had become one of the most recently industrialized regions of the world (Industrial System Research, 2003). Brazil, Russia, India, China, and South Africa (the BRICS States) are undergoing the process of industrialization (Sampath, 2016). There is a sizable literature on the factors facilitating industrial modernization and enterprise development (Lewis, 2003; Sampath, 2016).

Industrialization is the period of social and economic change that transforms a human group from an agrarian society into an industrial society. It involves an extensive re-organisation of an economy for the purpose of manufacturing (O'Sullivan & Sheffrin, 2003). Historically, industrialization is associated with an increase in polluting industries that were heavily dependent on fossil fuels. However, with the increasing focus on sustainable development and green industrial policy practices, industrialization increasingly includes technological leapfrogging, with direct investment in more advanced, cleaner technologies.

The reorganization of the economy has many unintended consequences, both economically and socially. As industrial workers' incomes rise, markets for consumer goods and services of all kinds tend to expand and provide a further stimulus to industrial investment and economic growth. Moreover, family structures tend to shift as extended families no longer tend to live together in one household, location, or place.

Between the early 1960s and 1990s, the Four Asian Tigers underwent rapid industrialization and maintained exceptionally high growth rates. The revolution was accompanied by a great deal of changes in the social structure, the main change being a transition from farm work to factory-related activities (Lee, 2016). This resulted in the creation of a class structure that differentiated the commoners from the well-off and the working category. It distorted the family system as most people moved into cities and left the farm areas, consequently playing a major role in the transmission of diseases. The place of women in the society then shifted from being home cares to employed workers hence reducing the number of children per household. Moreover, industrialization contributed to increased cases of child labor and thereafter, education systems (Lee, 2016).

Urbanization, exploitation (of labor and natural resources), change in family structure, and others were major social changes that accompanied industrialization. As of 2018, the "international development community" (World Bank, Organization for Economic Co-operation, and Development (OECD), many United Nations' departments – FAO, WHO, ILO and UNESCO) endorses development policies like water purification or primary education and co-operation amongst Third World communities (United Nations, 2015). Some members of the economic communities do not consider contemporary industrialization policies as being adequate to the Global South (Third World Countries) or beneficial in the longer term, with the perception that they may only create inefficient local industries unable to compete in the free-trade dominated political order which industrialization has fostered (Lee, 2016; United Nations, 2015). Environmentalism and Green politics may represent more intuitive reactions to industrial growth. Nevertheless, repeated examples in history of apparently successful industrialization (Britain, Soviet Union, South Korea, China, and others) may make conventional industrialization seem like an attractive or even natural path forward, especially as populations grow, consumerist expectations rise, and agricultural opportunities diminish.

The relationships between economic growth, employment, and poverty reduction are complex. Higher productivity, it is argued, may lead to lower employment (Melamed et al., 2011). There are differences across sectors, whereby manufacturing is less able than the tertiary sector to accommodate both increased productivity and employment opportunities; more than 40% of employees worldwide are "working poor" whose incomes fail to keep them and their families above the $2-a-day poverty line. There is also a phenomenon of deindustrialization, as in the former countries of the Union of Soviet Socialist Republics (USSR) transition to market economies, and the agriculture sector is often the key sector in absorbing the resultant unemployment (Melamed et al., 2011).

Principally, however, all of the developments of all the epochs, even from Adam and Eve in the Garden of Eden, through the period of fruit gathering to Industrial Revolution and the contemporary industrialization, the impact of human on the environment and vice versa have been humongous. The consequences, directly and indirectly, have been massive. The followings illustrate.

CONSEQUENCES OF INDUSTRIALIZATION ON THE ENVIRONMENT

Hitherto the period of the Industrial Revolution, literature on the subject were not common, but even then, there were records of domestic wastes; and pollution of water, land, and air were not common (Fleming & Knorr, 2006). The

Industrial Revolution brought a dramatic turnaround in respect of pollution resulting from the industrialization phenomena.

Levels of air pollution rose during the Industrial Revolution, sparking the first modern environmental laws to be passed in the mid-19th Century. The origins of the environmental movement lay in response to increasing levels of smoke pollution in the atmosphere during the period. The emergence of great factories and the associated immense growth in coal consumption gave rise to an unprecedented level of air pollution in industrial centers. After 1900, the large volume of industrial chemical discharges added to the growing loads of untreated human waste (Fleming & Knorr, 2006). The first large-scale, modern environmental laws came in the form of Britain's Alkali Acts, passed in 1863, to regulate the harmful air pollution (gaseous hydrochloric acid) given off by the Leblanc process used to produce soda ash. An Alkali inspector and four sub-inspectors were appointed to curb the pollution. The responsibilities of the inspectorate were gradually expanded, culminating in the Alkali Order 1958, which placed all major heavy industries that emitted smoke, grit, dust, and fumes under supervision (Fleming & Knorr, 2006).

The manufactured gas industry began in British cities in 1812-1820. The technique used produced highly toxic effluent that was dumped into sewers and rivers. The gas companies were repeatedly sued in nuisance lawsuits. They repeatedly lost and modified the worst practices. The City of London continually indicted gas companies in the 1820s for polluting the Thames and poisoning its fish. Finally, the parliament wrote company charters to regulate toxicity (Tomory, 2012). The industry reached the US around 1850, causing massive pollution and lawsuits (Tarr, 2014).

In industrial cities, local experts, and reformers, especially after 1890, took the lead in identifying environmental degradation and pollution and initiating grass-roots movements to demand and achieve reforms (Platt, 2005). Typically, the highest priority went to water and air pollution. The Coal Smoke Abatement Society was formed in Britain in 1898, making it one of the oldest environmental Non-Governmental Organizations. It was founded by artist Sir William Blake Richmond, who was frustrated with the pall cast by coal smoke. Although there were earlier pieces of legislation, the Public Health Act 1875 required all furnaces and fireplaces to consume smoke produced by them. It also provided for sanctions against factories that emitted large amounts of black smoke. The provisions of this law were extended in 1926 with the Smoke Abatement Act to include other emissions, such as soot, ash, and gritty particles, and to empower local authorities to impose regulations (William, 2014).

By and large, Kaveripakam (2022) alludes that industrial pollution takes many forms and has negative influence on the land, water, air, and the general environment, as well as causing sickness and death not only for humans but animals and devastating plant luxuriant growth. For instance, coal combustion, the combustion of fossil fuels such as petroleum, oil, natural gas, and chemical solvents used in the tanning and dyeing industries are the primary drivers of industrial pollution. Also, when fossil fuels such as coal and oil are used for power generation and transportation, they emit pollutants such as carbon dioxide, nitrogen, and sulfur dioxide, as well as fly ash, which damage the air, water, and land (Kaveripakam, 2022).

SOME WASTE AND POLLUTION DUE TO INDUSTRIALIZATION

Industries use raw materials, process them, produce finished products, and some by-products are thrown into the environment as industrial wastes in gas, liquid or solid, thus polluting the air, water, or soil. Industrial wastes may be classified as biodegradable and non-biodegradable:

1. Biodegradable wastes are generated by textile factories, food processing units, paper mills, cotton mills, and others.
2. Non-biodegradable wastes are generated by thermal power plants, iron and steel plants, plastic, and petrochemical plants and fertilizer industries, and others, posing severe threats to humankind.

Processes and Chemical Wastes

Process wastes produced during the washing and processing of raw materials may be organic or inorganic, but both are toxic to living organisms. Examples are organic wastes liberated from food processing units, distilleries, sugar mills, and others and inorganic process wastes generated by the caustic soda industry, the paint industry, the petroleum industry, iron and steel plants, and thermal power plants, and others. Fly ash from thermal power plants contaminates the atmospheric tract causing respiratory tract disorder. Fertilizer industries produce gypsum. Iron and steel plants have slag. Few industries might also cause thermal pollution and noise pollution, namely: Chemical wastes contain detergents, alkalis, acids, and other harmful substances produced by the industries, sugar mills, and others. The wastes are usually dumped into nearby water bodies like lakes, oceans, and others, altering the pH of biological and chemical oxygen demands. The aquatic animals and plants absorb chemical waste destroying trophic levels of ecosystems and food chains. However, the following have been adopted for the disposal of non-biodegradable industrial solid waste to control environmental pollution:

- The cement industry utilizes fly ash and slag from the steel industry.
- Wastes are subjected to proper treatment before its discharge.

FORMS OF INDUSTRIAL POLLUTION

Water Pollution

Industrial wastewaters contain a variety of organic and inorganic residues. They heavily pollute all rivers and water bodies. The discharge of toxic industrial wastes contains poisonous chemicals such as cyanide, cadmium, mercury, lead, arsenic, and chromium, which are highly poisonous. They make river water unfit for the use of humans, aquatic plants, and animals. Color-producing dyes change the color of water and decrease oxygen level, and thereby affecting aquatic life. In addition, acids and alkalis rapidly change water pH, affecting fish and other aquatic life forms.

Soil Pollution

The rapid growth of industries has resulted in the release of many industrial wastes containing toxic, acidic, and disastrous chemicals that are usually non-biodegradable. Solid waste from industries is dumped temporarily overland. During rains, heavy metals and toxic chemicals get washed down into the soil and pollute the same. It is mainly discharged from Pulp and Paper Mills, oil refineries, sugar factories, glass industries, drugs, and others. Industrial waste affects and alters the chemical and biological characteristics of soil, which finally enters the food chain, disturbs the biochemical processes, and eventually induces serious hazards to live organisms.

Air Pollution

Many industries like chemical plants, steel, fertilizers, sugar, and cement manufacturing units emit large amounts of smoke and pollutants like oxides of sulfur and nitrogen, lead particles, and chlorofluorocarbons pollute the air. For example, the emissions from the oil refinery at Mathura (Uttar Pradesh, India) and the numerous coal-burning industries at Agra (also in Uttar Pradesh, India) contain sulphur-dioxide that causes acid rain. In addition, many industries produce chemicals, some of which escape into the atmosphere causing air pollution.

Noise Pollution

Industrial activities cause noise pollution. Some familiar sources of noise pollution are road traffic, aircraft, trains, construction sites, factories, electronic and electrical equipment, and the bursting of firecrackers, among others.

CAUSES OF INDUSTRIAL POLLUTIONS

There could be many causes over space and time. However, they could be grouped as follows:

1. Unplanned Industrial Growth - In most countries, industrial development and urbanization have progressed unexpectedly. It is responsible for all types of pollution.
2. Use of Outdated Technology - Many industrial units are slow to adopt new technology to address pollution.
3. Use of Water for Industrial Processes - Most industrial units require large amount of water. Water used in different production processes comes into contact with harmful chemicals, heavy metals, biological waste, and others. After use, the water is dumped into water bodies causing water contamination.
4. Poor Implementation of Policies and Laws - Many developing countries do not have effective environmental policies and laws. Some firms do evade laws and policies.
5. Inefficient waste Disposal System - With unplanned and rapid industrial development, all countries generate industrial wastes quickly. All countries face the challenges of effectively disposing of industrial wastes, that is, including e-wastes, which contain toxic material.
6. Absence of Nation-Wide Pollution Tax - There is no nationwide pollution tax imposed in most countries. So, industrial pollution is very high; and it goes on unabated.

EFFECTS OF THE INDUSTRIAL POLLUTION

The effects could be enormous, widespread, and devastating (Embibe, 2022), namely:

- *Effect on Human Health* - Industrial pollution has been responsible for the contamination of water, air, and the natural environment. Moreover, it has affected the health of people. Industrial toxic waste is responsible for diseases like cancer, lung infection, asthma, and others.

- *Low Agricultural Productivity* - The toxic materials dumped by industrial units cause soil and groundwater contamination. It affects the fertility of the soil. In addition, the consumption of contaminated crops causes health problems in the long run.
- *Global Warming* - Global warming leads to rising water levels due to the melting of glaciers and constant threat of natural disasters like Tsunami and several storms. Due to global warming, many animals and fishes are becoming extinct.
- *Effect on Wildlife* - Industrial pollution and industrial activities have led to the destruction of the natural habitants of several animals and plants. As a result, many wildlife species face extinctions.
- *Depletion of Green Cover and Biodiversity* - The green cover helps balance temperature. Unfortunately, unregulated industrial activities have been responsible for the loss of green cover. Therefore, protecting areas and human health affected by rising temperatures due to global warming is essential.
- Industrial wastes discharged into water bodies contain many toxic substances that make water unfit for drinking and bathing. Moisture pollution also reduces the number of aquatic plants and animals due to the destruction of their habitat and nesting places.
- The wastewater released by factories and industries is rich in organic matter. It is rich in nutrients resulting in thick growth of algae and many other weeds like ragweed, skunk, buckthorn, and horsemint, and these plants cover the entire surface of some waters. The algae use so much oxygen; hence aquatic animals and other plants die due to lack of oxygen.
- When industries release the mercury, it contaminates with water, and it is used for drinking by human and animals; it causes numbness of lips, tongue, and limbs. Also, it leads to blurred vision and mental disorders.

TOWARDS CONTROLLING THE INDUSTRIAL POLLUTION

The foregoing has identified various forms of the industrial waste and their sources, and their consequences. There are obvious needs to control the pollution. For instance, to inspect the air pollution by industrial plants and chimney wastes, several measures are being adopted to remove particulate matter and gaseous pollutants from the scraps. The most common equipment used are cyclone collectors, electrostatic precipitators, bag filters, and scrubbers (Kaveripakam, 2022).

- Cyclone Collectors - Removing particulates through vortex separation without filters from the liquid stream, gas, or air is called Cyclonic separation. Cyclone collectors are instruments used to trap the particles like dust produced by wood and cement factories and others.
- Electrostatic Precipitators - These are charging the dust by the applications of high voltage electricity. The particles finally settle down. These are safe and simple to operate.
- Bag Filters - It consists of filter bags made up of bag materials like cotton, glass wool, Teflon, ceramic fiber or polyester, and others. The dust-laden gas pass through it. As a result, the dust gets filtered, and clean air escapes. The dust collected is removed by shaking the bag periodically.
- Scrubbers - Scrubbers can remove the harmful gas released by different factories by spraying cool water in a scrubber device. The gaseous pollutants observe inappropriate liquid to bring the pollutants from gaseous to liquid or solid-state - the different types of scrubbers used are spray type, venturi scrubbers, impingement scrubbers, and others.

The pollution due to industries may be eliminated by the use of electricity in place of coal fuel. The extent of air pollution can be reduced by self-cleaning processes of the air. It can increase vegetation in the nearby locality and provide Green Belt between residential and industrial areas (Kaveripakam, 2022).

In some similar ways, solid, liquid and noise pollutions need to be prevented and or controlled. There are a number of rudimentary and scientific methods in literature (Aderogba, 2013; Broadberry & Gupta, 2005; Goudie, 2000; Kaveripakam, 2022; McCloy, 2019; Sampath, 2016; Tarr, 2014).

Above all, and in all circumstances, humans are persistently requiring a sustainable environment in the processes of procuring his food, water, clothing, pleasure and recreation, transport and communication, industrialization, urban development, and others, but these can be done in conformity with Education for Sustainable Development and Education for Sustainable Environment.

EDUCATION FOR A SUSTAINABLE ENVIRONMENT

Education for sustainable development (ESD) will promote the development of the knowledge, skills, understanding, values, and actions required to create a sustainable world, which will ensure environmental protection and conservation, promotes social equity, and encourages economic sustainability. It is an "interdisciplinary learning methodology covering the integrated social, economic, and environmental aspects of formal and informal curriculum" (Higher

Education Academy, 2022; Manno & Fix, 2019, p. 7). It is an educational ideology that recognizes modern-day environmental challenges and seeks to define new ways to adjust to a changing biosphere, as well as engage individuals to address societal issues that come with them (Jucker & Mathar, 2014). In the International Encyclopedia of Education, this approach to education is seen as an attempt to "shift consciousness toward an ethics of life-giving relationships that respects the interconnectedness of man to his natural world" in order to equip future members of society with environmental awareness and a sense of responsibility to sustainability (Peterson et al., 2014). The Brundtland Commission defined sustainable development as meeting the needs of the present generation without putting at risk the capacity of generations to come to meeting their requirements (United Nations World Commission on Environment and Development, 1987). There have been two more global activities since then. These were the United Nations Conference on Environment and Development in 1992 (Rio Earth Summit) and the United Nations World Commission on Environment and Development in 2016. According to the UNESCO, the education for sustainable development involves:

> "integrating key sustainable development issues into teaching and learning. This may include, for example, instruction about climate change, disaster risk reduction, biodiversity, poverty reduction, and sustainable consumption. It also requires participatory teaching and learning methods that motivate and empower learners to change their behaviours and act for sustainable development. ESD consequently promotes competencies like critical thinking, imagining future scenarios and making decisions in a collaborative way" (Marope et al., 2015 & UNESCO, 2013).

The Thessaloniki Declaration, presented at the "International Conference on Environment and Society: Education and Public Awareness for Sustainability" by UNESCO and the Government of Greece (December 1997), highlights the importance of sustainability not only with regard to the natural environment but also with poverty, health, food security, democracy, human rights, and peace (Nikolopoulou, 2010).

The concept of ESD was born from the need for education to generally address the growing environmental challenges facing the planet Earth. Sustainability in higher education is not only limited to embedding intended learning outcomes about sustainable development into the curriculum of higher educational institutions. However, a sustainable campus should integrate the educational and managerial aspects of sustainable development along with its three dimensions (environmental, economic, and social responsibility) into its different practices (Adel & Mahrous, 2018; University of California, 2021).

Education for sustainable development (ESD) encourages changes in knowledge, skills, values, and attitudes to enable a more sustainable and just society for all (UNESCO, 2013). It will empower and equip current and future generations to meet their needs using a balanced and integrated approach to the economic, social, and environmental dimensions of sustainable development (UNESCO, 2018). The term ESD is used internationally and by the United Nations.

Education must be strengthened in all agendas, programs, and activities that promote sustainable development. Sustainable development must be integrated into education, and education must be integrated into sustainable development. The ESD promotes the integration of these critical sustainability issues, in local and global contexts, into the curriculum to prepare learners to understand and respond to the changing world. The ESD aims to produce learning outcomes that include core competencies such as critical and systematic thinking, collaborative decision-making, and taking responsibility for the present and future generations. Since traditional single-directional delivery of knowledge is not sufficient to inspire learners to function as responsible citizens, ESD entails rethinking the learning environment, physical and virtual. The learning environment must adapt and apply a whole-institution approach to embed the philosophy of sustainable development. Building the capacity of educators and policy support at international, regional, national, and local levels helps drive changes in learning institutions. Empowered youth and local communities interacting with education institutions become key actors in advancing sustainable development (UNESCO, 2018).

Agenda 21 of the Sustainable Development Goal was the first international document that identified education as an essential tool for achieving sustainable development and highlighted areas of action for education (Adel & Mahrous, 2018). ESD is a component of measurement in an indicator for Sustainable Development Goal 12 (SDG) for responsible consumption and production. SDG 12 has 11 targets. Target 12.8 says that by 2030, ensure that people everywhere have the relevant information and awareness for sustainable development and lifestyles in harmony with nature (United Nations, 2017).

UNESCO (2021) Education for Sustainable Development (ESD), as earlier noted, empowers learners of all ages with the knowledge, skills, values, and attitudes to address the interconnected global challenges the world is facing, including climate change, environmental degradation, loss of biodiversity, poverty, and inequality. Learning must prepare students and learners of all ages to find solutions for the challenges of today and the future. Education should be transformative and allow humankind to make informed decisions and take individual and collective action to change the societies and care for the planet Earth.

Education for Sustainable Development is recognized as an integral element of Sustainable Development Goal (SDG) 4 on quality education and a key enabler of all other SDGs. As the lead UN agency on ESD, UNESCO is responsible for the coordination of the ESD for 2030 Framework (UNESCO, 2021). Transformative education involves the teaching and learning geared to motivate and empower learners to make informed decisions and actions at the individual, community, and global levels.

From November twenty ninth to December first, December 2021, UNESCO, and the Asia-Pacific Centre of Education for International Understanding (APCEIU) brought together experts in education for sustainable development, global citizenship education, and education for health and well-being to discuss good practices, progress, monitoring, and mainstreaming transformative education towards Sustainable Development Goal target 4.7. 1600 policy makers, education practitioners, experts, and young people actively participated in the Forum to share their experiences and knowledge: and suggested ways forward. But what is Environmental Education?

Environmental Education

The United States Environmental Protection Agency (EPA) (2021), diligently explains and outlines the content: It is a process that allows individuals to explore environmental issues, engage in problem-solving, and take action to improve the environment. As a result, individuals develop a deeper understanding of environmental issues and have the skills to make informed and responsible decisions. The components are:

- Awareness and sensitivity to the environment and environmental challenges.
- Knowledge and understanding of the environment and environmental challenges.
- Attitudes of concern for the environment and motivation to improve or maintain environmental quality.
- Skills to identify and help resolve environmental challenges; and
- Participation in activities that lead to the resolution of environmental challenges.

It does not advocate a particular viewpoint or course of action. Rather, it teaches individuals how to weigh various sides of an issue through critical thinking, and it enhances their own problem-solving and decision-making skills.

The National Environmental Education Act of 1990 requires EPA to provide national leadership to increase environmental literacy. EPA established the Office of Environmental Education to implement this programs.

It is important to distinguish between Environmental Education and Environmental Information: Environmental Education is more than information about the environment. It increases public awareness and knowledge of environmental issues, teaches individuals critical-thinking, enhances individuals' problem-solving and decision-making skills, but it does not advocate any particular viewpoint. Conversely, Environmental Information provides facts or opinions about environmental issues, but would not necessarily teach individuals critical thinking, nor enhance individuals' problem-solving and decision-making skills, but may advocate a particular viewpoint.

CONCLUSION

The human impacts on the environment over the years have been humongous. As a result, the natural environment is limited, and it will keep on diminishing. The chapter has provided an overview of industrial revolutions; gave an insight into what is a 'natural' environment and how it has been interfered with directly and inadvertently by the processes of industrialization over space and time. It dwelt on the consequences of man-environment interaction and outlined the position of Environmental Education towards a sustainable environment.

Environment to change and the role of human in bringing the changes about are circularly causal and continual – developments over space and time. What will 'remain' of any environment must be sustainable for lives - "development that meets the needs of the present without compromising the ability of future generations to meet their own needs." As a matter of policy, therefore, governments and regions should:

- Re-examine the critical issues of environment and development and formulate innovative, concrete, and realistic action proposals to deal with them, that is, in congruence with the United Nations.
- Propose short-term and long-term environmental strategies for achieving sustainable development to at least the next 50 years.
- Recommend ways in which concern for the environment may be translated into greater co-operation among developing countries and between countries at different stages of economic and social development and lead to the achievement of common and mutually supportive objectives that will take account of the interrelationships between people, resources, environment, and development.
- Consider ways and means by which the international community can deal more effectively with environmental concerns in the light of the other recommendations of the United Nations.

- Help to define shared perceptions of long-term environmental issues and of the appropriate efforts needed to deal successfully with the problems of protecting and enhancing the environment, a long-term agenda for action in the nearest future, and aspirational goals for the world community, considering the relevant resolutions of the session of a special charter of the Governing Council of the United Nations.
- Strengthen international cooperation on environment and development, assess and propose new forms of cooperation that would break out of existing patterns and influence policies and events in the direction of needed change; and
- Raise the level of understanding and commitment to action on the part of individuals, voluntary organizations, businesses, institutes, and governments.

United Nations General Assembly focuses its attention on the areas of population, food security, loss of species and genetic resources, energy, industry, and human settlements - realizing that all of these are connected and cannot be treated in isolation. Governments and regional organizations would have to work in collaboration with the United Nations General Assembly in these respects and give cognizance of industrial development in any region of the 21st Century.

It is in the "environment" where humans live; and "development" is what all embark on in attempting to improve the lots within their places of abode. The two are inseparable. Thus, the 21st Century environment should also include social and political atmospheres and circumstances, as well as recognizing that development is not just about how developing countries can ameliorate their situation, but what the entire world, including developed countries, can do to ameliorate their common situations.

REFERENCES

Adel, H. M. & Mahrous, A. A. (2018). Sustainability communication and evaluation: A practice-based case study on British-Egyptian universities value-chain. Bristol Business School, University of the West of England: Proceedings of the 32nd Annual International Conference of The British Academy of Management (BAM) 2018: Driving Productivity in Uncertain and Challenging Times. September 4 – September 6, 2018.

Aderogba, K. A. & Bankole, M. O. (2016). Unsustainable exploitation of the Guinea Savannah, South of Sahara: Issues, policies and programs for sustainability, *Environment and Sustainable Development in the 21St Century, Ife Social Sciences Review Journal, (Special Issue)* pp. 78-95.

Aderogba, K. A. (2013). Polymer waste and management in cities and towns of Africa and sustainable environment: Nigerian experience. In Maurice N. Amutabi (Ed.) *Environment, Power and Development in Africa,* A publication of the Catholic University of Eastern Africa (CUEA), Nairobi, Kenya, pp. 260-274.

Aderogba, K. A. (2016). Growing slums, environmental degradation, and sustainable development in Lakowe-Ajah communities in Lagos State. *LASU Journal of Public Administration and Management* 2(2), 64-82.

Aderogba, K. A. (2021). Man-environment interaction in the rainforests and sustainable development: Practical implications for Adult Education. In M. V. Alfred, P. A. Robinson, & E. A. Roumell (Eds.) *Advancing the Global Agenda for Human Rights, Vulnerable Populations, and, Environmental Sustainability: Adult Education as a Strategic Planner.* Information Age Publishing, pp. 67-82.

Aderogba, K. A., Bankole, M. O. & Komolafe, B. O. (2021). Weather and climate, impacts on livelihood and sustainability in Ijebu-Ode Region. In A. S. Gbadegesin, F. O. Oderinde, G. O. Enaruvbe and H. D. Olaniran (Eds.) *Towards Achieving a Sustainable Environment: Perspectives from Geography, Ecology and Environmental Science. A Festschrift for Professor Albert Orodena Aweto.* Ibadan: Department of Geography, Faculty of The Social Sciences, University of Ibadan, Nigeria, pp. 224-252.

Ask Media Group. (2021 June 10). Earth's Five Types of Natural Environments. Ask Media Group. https://www.reference.com/science/types-environments-1-327160a5c260ec1e.

Awobajo, K. A. (1980). Effect of Continuous Cropping on Soil Property in Ilishan, Remo, Ogun State, Nigeria, Unpublished B. Sc. Dissertation, Department of Geography, Faculty of Science, University of Ibadan, Ibadan.

Berg, M. & Hudson, P. (1992). Rehabilitating the Industrial Revolution. *The Economic History Review, 45* (1), 24-50.

Bluhdorn, H. (2017). Post-capitalism, post-growth, post-consumerism? Eco-political hopes beyond sustainability". *Global Discourse, 7*(1), 42-61.

Bond, E., Gingerich, S., Archer-Antonsen, O., Purcell, L. & Macklem, E. (2003 February 27). The Industrial Revolution – Innovations. Industrialrevolution.sea.ca. Retrieved 25 January 2022.

Broadberry, S. & Gupta, B. (2005). Cotton Textiles and the Great Divergence: Lancashire, India, and Shifting Competitive Advantage, 1600–1850. Department of Economics, International Institute of Social History. University of Warwick.

Broadberry, S. N., Guan, H. & Li, D. D. (2017 April 1). *China, Europe, and the Great Divergence: A Study in Historical National Accounting, 980–1850.* CEPR Discussion Paper.

Buchheim, C. (1994). Industrielle Revolutionen. Langfristige Wirtschaftsentwicklung in Großbritannien, Europa und in Übersee, München, pp. 11-104.

Crafts, N. (2005). The first industrial revolution: Resolving the slow growth/rapid industrialization paradox. *Journal of the European Economic Association* 3(2-3), 525-534.

Crafts, N. (2011 April 1). Explaining the first Industrial Revolution: Two views. *European Review of Economic History, 15*(1): 15368. doi:10.1017/S1361491610000201.

Deane, P. (1973). The Industrial Revolution in Great Britain. In C. M. Cipolla (Ed.) *The Fontana Economic History of Europe: The Emergence of industrial societies* 4(2), 161-174.

Embibe (For Schools, Competitive Exams and Beyond). (2022). Industrial Pollution: Definition, Causes, Effects, Prevention. https://www.embibe.com/exams/industrial-pollution/.

Esposito, J. L. (Ed.) (2004). The Islamic World: Past and Present. Vol. 1: Abba - Hist. Oxford University Press, p. 174.

Feinstein, C. (1998 September). Pessimism Perpetuated: Real Wages and the Standard of Living in Britain during and after the Industrial Revolution. *Journal of Economic History, 58*(3), 625-58.

Fleming, J. R. & Knorr, B. R. (2006). History of the Clean Air Act. *American Meteorological Society.* Retrieved 18 February 2022.

Goudie, A. (2000). *The Human Impact on the Natural Environment.* MIT Press. pp. 203-239.

Griffin, E. (2010). *Short History of the British Industrial Revolution.* Palgrave, pp.35-79.

Higher Education Academy. (2022). Education for sustainable development. UNESCO/Higher Education Academy. https://en.unesco.org/themes/education-sustainable-development.

Hudson, P. (1992). The Industrial Revolution. Edward Arnold, p. 11.

Ijezie, L. E. (2021 March 31). Vocation of humanity in Genesis 2-3 and its implications for Eco-Theology in Africa. *European Journal of Theology and Philosophy 1*(2), 1-6. DOI :10.24018/ejtheology.2021.1.2.10.

Industrial System Research. (2003). Industry & Enterprise: An International Survey of Modernisation & Development, ISM/Google Books, revised 2nd edition, 127 pp.

Jérôme-Adolphe, B. (1837). Histoire de l'économie politique en Europe depuis *les anciens jusqu'à nos jours*, 1(3), 383.

Johnson, D. L., Ambrose, S. H., Bassett, T. J., Bowen, M. L. Crummey, D. E., Isaacson, J. S., Johnson, D. N., Lamb, P., Saul, M. & Winter-Nelson, A. E. (1997). Meanings of Environmental Terms. *Journal of Environmental Quality, 26*(3), 581-589.

Jones, E. (2003). The European Miracle: Environments, Economics and Geopolitics in the History of Europe and Asia, (3 Ed.) Cambridge, p. 135 -166.

Kaveripakam, S. (2022 February 1). Industrial Pollution: Definition, Causes, Effects, Prevention. Embibe. https://www.embibe.com/exams/industrial-pollution/.

Klein, E. (2021 September 19). Opinion: The Economic Mistake the Left Is Finally Confronting. The New York Times.

Landes, D. (1999). *The Wealth and Poverty of Nations.* W.W. Norton & Company.

Landes, D. S. (1969). The Unbound Prometheus: Technological Change and Industrial Development in Western Europe from 1750 to the Present. Press Syndicate of the University of Cambridge, pp. 15-40.

Lee, R. (2016 September 26). Industrialization and Exploitation. Medium. Archived from the original on 3 July 2020. Retrieved 30 June 2020.

Lucas, R. E. (2002). Lectures on Economic Growth. Harvard University Press, pp. 109-110.

Macmillan, K. (2020 April 11). Human-Environment Interaction Examples. https://az greenmagazine.com/human-environment-interaction/#:~:text=Human-environment%20 interaction%20is%20the%20interrelation%20between%20humans%20 and,there%20are% 20three%20basic%20types%20of%20this%20interrelation%3A.

Manno, J. & Fix, A. (2019 July 29). Environmental Sustainability and Sustainable Development. International Studies Association and Oxford University Press. https://oxfordre.com/ internationalstudies/view/10.1093/acrefore/9780190846626.001.0001/acrefore-9780190 846626-e-169#:~:text=Environmental%20sustainability%20as%20a%20topic%20in%20 international%20studies,of%20economic%20development%2C%20environmental%20 protection%2C% 20and%20social%20progres.

Marope, P.T.M., Chakroun, B. & Holmes, K. P. (2015). Unleashing the Potential: Transforming Technical and Vocational Education and Training. UNESCO, pp. 9, 23, 25-26.

McCloy, J. (2019 October 17). Caring for the environment: 8 Reasons to protect and sustain earth. Greencoast. https://greencoast.org/caring-for-the-environment/.

Melamed, C., Hartwig, R. & Grant, U. (2011). Jobs, growth, and poverty: what do we know, what don't we know, what should we know? Archived 20 May 2011 at the Wayback Machine. Overseas Development Institute.

Mendels, P. & & Kridte, K. (2001). The Great Divergence: China, Europe, and the Making of the Modern World Economy (Princeton Economic History of the Western World). Princeton University Press, 118-217.

Milward, A. S. & Berrick, S. T. (1977). *The Development of the Economies of Continental Europe 1850–1914.* Harvard University Press, p. 77.

Milward, A. S. & Saul, S. B. (1973). *The Economic Development of Continental Europe 1780-1870* Harvard University Press, pp. 132-142.

Nikolopoulou, A., Abraham, T., & Mirbagheri, F. (2010). *Education for sustainable development: Challenges, strategies, and practices in a globalizing world.* SAGE Publications Inc. https://doi.org/10.4135/9788132108023

Ogilvie, S. (2008). Protoindustrialization. In S. Durlauf, B. Steven & L. Blume (Eds.). *The New Palgrave Dictionary of Economics.* Palgrave Macmillan (6), 711-714.

O'Sullivan, A. & Sheffrin, S. M. (2003). Economics: Principles in Action. Pearson Prentice Hall, p. 472.

Peterson, P. L., Baker, E. L. & McGaw, B. (2014). International encyclopedia of education. (3rd Ed.). Elsevier Science, pp. 198-202.

Platt, H. L. (2005). Shock cities: the environmental transformation and reform of Manchester and Chicago excerpt.

Pollard, S. (1981). Peaceful Conquest. The Industrialisation of Europe 1760–1970, Oxford, pp. 117-156.

Purvis, B., Mao, Y. & Robinson, D. (2019). Three pillars of sustainability: in search of conceptual origins. *Sustainability Science, 14*(3), 681-695. doi:10.1007/s11625-018-0627-5.

Raskin, P. (2002). Great Transition: The promise and lure of the times ahead. Global Scenario Group. Stockholm Environment Institute. OCLC 49987854.

Ray, I. (2011). Bengal Industries and the British Industrial Revolution (1757-1857). Routledge, pp. 7-10.

Reference.com. (2021 June 10). Earth's Five Types of Natural Environments. https://www. reference.com/science/types-environments-1-327160a5c260ec1e.

Reisman, G. (1998). Capitalism: A complete understanding of the nature and value of human economic life. Jameson Books, p. 127.

Rider, C. (Ed.) (2007). *Encyclopedia of the Age of the Industrial Revolution 1700–1920*, pp xiii - xxxv.

Sampath, P. G. (2016). Sustainable Industrialization in Africa: Toward a New Development Agenda. *Sustainable Industrialization in Africa.* Springer, pp. 1-19.

Smith, N. (2022 February 3). A New Industrialist roundup. *Noahpinion.* Retrieved 5 February 2022.

Symons, D. (1979). *The Evolution of Human Sexuality.* Oxford University Press. p. 31.

Szreter, S. & Mooney, G. (1998 February). Urbanization, Mortality, and the Standard of Living Debate: New Estimates of the Expectation of Life at Birth in Nineteenth-Century British Cities. *The Economic History Review, 51*(1), 81-112.

Tarr, J. A. (2014). Toxic Legacy: The Environmental Impact of the Manufactured Gas Industry in the United States. *Technology and culture 55*(1), 107-47.

The University of California, Los Angeles (UCLA). (2021). What is Sustainability? UCLA. https://www.sustain.ucla.edu/what-is-sustainability/.

Tomory, L. (2012). The Environmental History of the Early British Gas Industry, 1812–1830. *Environmental history 17*(1), 29-54.

Tong, J. T. (2016). Finance and Society in 21st Century China: Chinese Culture Versus Western Markets. CRC Press, p. 151.

United Nations Educational, Scientific and Cultural Organization (UNESCO). (2021). Education for Sustainable Development. UNESCO. https://en.unesco.org/themes /education-sustainable-development.

United Nations Educational, Scientific and Cultural Organization (UNESCO). (2018). Issues and trends in education for sustainable development. UNESCO, p. 7.

United Nations Educational, Scientific and Cultural Organization (UNESCO). (2013 May 10). *Education for Sustainable Development.* UNESCO.

United Nations Educational, Scientific and Cultural Organization. (UNESCO). (2015). *Sustainable Development.* UNESCO. https://en.unesco.org/themes/education-sustainable-development/what-is-esd/sd.

United Nations World Commission on Environment and Development (1987). Report of the World Commission on Environment and Development: Our Common Future. Transmitted to the General Assembly as an Annex to document A/42/427 – Development and International Co-operation: Environment.

United Nations. (2015). United Nations Millennium Development Goals: News on Millennium Development Goals. United Nations. https://www.un.org/millennium goals/.

United Nations. (2017). Envision2030 Goal 12: Responsible Consumption and Production. United Nations. https://www.un.org/development/desa/disabilities/envision2030-goal12.html.

United Nations. (2017a). Resolution adopted by the General Assembly on 6 July 2017, Work of the Statistical Commission pertaining to the 2030 Agenda for Sustainable Development (A/RES/71/313). United Nations.

United Nations. (2017b). World population projected to reach 9.8 billion in 2050, and 11.2 billion in 2100. Department of Economics and Social Affairs, United Nations. https://www. un. Org /en/desa/world-population-projected-reach-98-billion-2050-and-112-billion-2100#:~:text= The %20current%20world%20population%20of%207.6%20billion%20is,a%20new%20 United%20Nations%20report %20being%20launched%20today.

United States Environmental Protection Agency (USEPA). (2021a August 18). What is Environmental Education? UNEPA. https://www.epa.gov/education/what-environmental-education.

United States Environmental Protection Agency (USEPA). (2021b). *Our Built and Natural Environments: A Technical Review of the Interactions Between Land Use, Transportation, and Environmental Quality* (2nd Edition). United States Environmental Protection Agency.

Wrigley, E. A. (2018). Reconsidering the Industrial Revolution: England and Wales. *Journal of Interdisciplinary History 49(*01), 9-42.

Zondervan. (2011). *Holy Bible: The Book of Genesis. New International Version* (NIV), Zondervan.

Chapter 5

The Last Decade of the 20th Century

By Veronica Woodard & Twila Sterling-Guillory

INTRODUCTION

Ten years ago, former Vice President Al Gore (Research by Cook, 2016) organized a documentary called, "An Inconvenient Truth," which opened in cinemas in the United States. This documentary depicts the threat climate change has no doubt made on this country. Its focus was to raise public awareness and change people's behavior to decrease greenhouse gas emissions. The documentary was so impactful that it won two Academy Awards, with Vice President Gore winning the 2007 Nobel Peace Prize for his determination to communicate human-induced climate change. Abbreviated AIT for short it has been reported that this documentary is the eleventh highest-grossing documentary in the United States. Texan climatologist Steve Quiring (2016) stated it this way: "AIT has had a much greater impact on public opinion and public awareness of global climate change than any scientific paper or report."

This chapter will offer an analysis of several different subject areas in the last ten years of the 20th century. Some of the areas will include human effects of climate change, pandemic illnesses such as, Auto Immune Deficiency Syndrome (AIDS), and Ebola, as well as greenhouse gases. Additionally, the digital transformation of health care will be addressed in relation to the last decade of the 20th century.

CLIMATE CHANGE

Individuals who have been tagged as climate communicators have agreed that their awareness of the issue of climate change/global warming began after watching the documentary, An Inconvenient Truth (AIT), by former Vice President Al Gore. Several studies have experimentally conducted public surveys to assess the impact viewing the documentary had on the public's view of climate change (Cook, 2016). The Pew Research Center for People & the Press was one of the companies which performed this survey. Prior to the documentary's release, 41% of Americans believed global warming was due to human activity. In the months following the release of the documentary, this belief increased to 50 percent.

Adding proof to this belief, surveys of motion-picture fans and students realized that watching AIT increased knowledge regarding the causes of global warming and willingness to decrease greenhouse gases. The increased knowledge did not lead to a change in human behavior. Confirmation was received by a follow-up survey conducted a month later.

Public attitude towards climate change was analyzed, noting a substantial relationship between the mentioning of AIT by the media and public perception concerning the urgency of climate change. Another significant factor related to this study found that polarization about climate change decreased following the release of AIT. This defeated the claim that Al Gore polarized the climate debate. Leaders from both the Democratic and Republican parties were credited with being the major drivers of public polarization on climate change.

While research accentuates the difficulty associated with confronting public awareness campaigns, this did not apply to AIT. Three profound factors were attributed to the AIT documentary: 1) AIT was successful in raising public awareness of climate change 2) The AIT documentary increased willingness to change behavior, and 3) In some cases, the behavior did change (Cook, 2016).

PANDEMICS

Auto Immune Deficiency Syndrome (AIDS)

Laboratory and clinical science appeared to have mastered epidemic infectious diseases in the late twentieth century. Several societies were confronted with a critical new infectious disease in the early 1980s. This disease was categorized as a heterosexual epidemic in Africa. Combatting this disease was difficult for multiple reasons. First, there were no pre-existing information networks specific to the new disease. Second, there were no standard procedures to treat the new infectious disease. Third, there were no established areas of expertise. The disease was linked to so-called 'deviant minorities' such as blacks, gay individuals, and drug users. In the early days of AIDS, an abundance of social hostility was the common reaction.

Historian, Roy Porter (1986), was the author of an editorial in the *British Medical Journal* titled, "History Says No to the Policeman's Response to AIDS." This document provided a compelling argument against adding AIDS to the list of notifiable diseases, using the historical precedent of the Contagious Diseases Acts. His specific objection was stated as follows:

"Historical precedent says 'no.' For unlike casually contagious diseases, sexually transmitted diseases constitute a special case in which the direct methods of the law to have been tried, found wanting, and abandoned.... Desperate diseases may require desperate remedies. Faced with the enormity of the suffering AIDS will inflict, humanity demands that we at least consider draconian measures such as compulsory screening for suspected virus carriers and further steps to protect others. Experience suggests, however, that this would be unwise" ... (*British Medical Journal*, 20, pp. 1589-1590).

Unlike casually contagious diseases, sexually transmitted diseases make up an unusual case where the direct methods of the law have been involved. It's been phrased, that desperate diseases sometimes require desperate remedies. AIDS inflicted enormous suffering on countless individuals. Because of this, it was demanded by humanity that extreme measures be cared out, including compulsory screening for those who were suspected of carrying the virus, along with steps to protect individuals from contracting the virus.

AIDS has been referred to as a disease that is not fully understood and difficult to treat, despite all the advances which have been made in the late 20th century (Weiss, 1997). AIDS has also been metaphorically known to "make cells die." The enemy relating to AIDS is the HIV virus which penetrates the helper T cells of the immune system. This turns the body "against itself," via altering the body's mutation or viral infection. AIDS is a "staged" disease which can develop from HIV seropositive benign tumors (latency) and that can lead to death. Individuals who have contracted AIDS have been identified as members of a certain "risk group." These same persons have also been ostracized.

Now a part of Western biomedical and popular culture, AIDS is one of the diseases that has been kept under scrutiny. In the past, numerous sociological studies have attempted to describe and analyze the many aspects of AIDS. Unfortunately, a stigma has been attached to the disease. Katz et al. (1987) identified AIDS as "the only patient group that seems to be devalued, blamed, and feared more than cancer patients by nonpatient are AIDS sufferers,"(Katz et al., 1987 p.458).

AIDS is now well into its second decade. It has had a colossal influence on science and society. It's recognized that AIDS has transformed many "minority" populations such as homosexuals, prostitutes, and drug users, only to name a few, into "risk sufferers." This has been interpreted by some as emphasizing and projecting a politics of stigma and marginality (Parker et al., 1991; Treichler, 1987, 1992; Turner et al., 1989). Our life has been shaped by AIDS, determining our perspective on how we see certain things. One such example is how we should engage in sexual intercourse. It has redefined what is safe and what is risky (Gagnon, 1989; Reiss & Leik, 1989). Boxer and Cohler (1989) phrased it this way: "It redesigns the cultural life course of those subjected it." Bolton (1989) claims, "AIDS is a phenomenon that afflicts societies and cultures while HIV undermines the health of individuals."

AIDS has been classified as an epidemic of morality and stigma. Patients who have been diagnosed with this disease are more than members of a risk group; they are also at times referred to as patients who are "polluted carriers" of an evil disease who should be avoided and punished for their misconduct, deviance, and delinquency. According to Douglas (1966), the emphasis on "pollution" suggests that beliefs about pollution and contagion express, maintain, and reinforce ideas of social morality and order. Specifically, this is what was stated:

"Pollution rules in essence prohibit physical contact. They to be applied to products of human physiology; thus, they regulate contact with blood, excreta, vomit, hair clippings. It seems that physiological pollutions become important as symbolic expressions of other undesirable contacts which would have repercussions on the structure of social ideas." (Douglas, 1975 (1968):55)

Various students, nurses, and doctors have admitted to depicting AIDS patients as individuals restricted by a social frame as opposed to the blurred fragments of the boundaries of their body. Repeatedly, research on the AIDS disease has revealed the way in which fear, discriminatory attitudes, and beliefs regarding pollution have impeded and, at time, paralyzed efforts to provide treatment for those afflicted with AIDS (Bayer, 1989; Marshall et al., 1990; Sabatier, 1988).

Frankenberg (1994) connotated there is a link between AIDS and select "minority" groups, which motivates the so-called "culture of AIDS". Furthermore, in his study, he also found no reference to a "local culture of AIDS," risk groups, or any other demographic aspect of the disease. Additionally, Frankenberg suggests that metaphors of AIDS extend "beyond culture;" those metaphors can be both local and universal.

AIDS has been characterized as drawing particular attention to "risk groups; conversely, it is also a global pandemic moving across national boundaries, which has been known to afflict "all" societies and cultures. Although the history of AIDS is in its second decade, two phases have been identified. First, the arrival of AIDS was accompanied by the shocking realization of the high risk of contamination along with the tumultuous reaction and stigma of marginalization regarding "risk groups." The second phase that we are now experiencing is bringing us to the realization that AIDS patients are all around us. Whether they are our neighbors, our relatives, heterosexual individuals classified as middle-class financially, individuals who received contaminated blood transfusions, or a myriad of other factors which lead to the diagnosis of AIDS, we have concluded that AIDS can afflict anyone, anywhere, at any time. The end result is AIDS is a cultural and universal.

Ebola

The Ebola virus was discovered in 1976 in a village along the Ebola River in the Democratic Republic of the Congo (DRC) (Sullivan, 2021). It is a highly contagious virus with the prominent symptoms being severe fever and internal bleeding. The virus is spread via contact with bodily fluids. Treatment must be prompt as soon as the virus is detected. Without immediate treatment once Ebola has been detected, the virus can spread quickly throughout families and communities. The death rate of people who contract Ebola is approximately 50 percent (Sullivan, 2021). Ebola is one of the deadliest infectious disease threats known to non-gender specific (Zeng, 2021). There is a high level of the virus which remains lurking in the body even after a patient has recovered. It is unknown where this hiding place remains in the body.

In an effort to better understand where the Ebola virus "hides" in the body of survivors, triggering recurrent infections, 36 rhesus monkeys were examined that had been treated for Ebola with monoclonal antibody therapy. Specifically, this is a type of treatment that aids the immune system in attacking an infection. The monkeys were regarded as fully recovered with no symptoms of infection or detectable virus in their blood (Zeng, 2021).

In a detailed microscopic examination of tissue from different organs, it was discovered that approximately 20% of the monkeys who had recovered from Ebola still had visible Ebola virus located exclusively in the ventricular system of the brain. The ventricular region of the brain produces, circulates, and stores cerebrospinal fluid, protecting, supplying nutrients to, and removing waste products from the brain (Zeng, 2021).

Although the monkeys were asymptomatic at the beginning of the study, two of the monkeys observed who were included in the study, developed Ebola symptoms before dying 30 and 39 days following their initial infection, respectively. The findings from this study suggests that the Ebola virus can hide dormant in the survivors' brains even after treatment for the disease. It was also determined that the virus could reactivate and cause fatal infections later.

There were three Ebola outbreaks in Africa In 2021 (Zeng, 2021). Each of the outbreaks was linked to previously infected survivors. Ebola also reemerged in Guinea that same year, linked to a survivor of the 2013-2016

Ebola outbreak (Zeng, 2021). High fever is one of several symptoms associated with the Ebola virus. Following the incubation period, the virus begins to produce symptoms of a high fever. High fever mimics various other diseases, including yellow fever or malaria. Both are common in the region where Ebola exists. For this reason, people may have accidentally been exposed to the virus when providing care to family members.

Ebola fever rarely breaks. When it does, the patient may gain only temporary relief. The reason this occurs is due to the body producing more pyrogens to fight the virus by turning up the body temperature of the infected individual.

Headaches and joint and muscle aches are also symptoms of infection with Ebola. When an individual develops an infection, the immune system receives a signal from the brain to produce antibodies to fight the disease, creating white blood cells from T-cells. The object of this function is to kill the virus. However, healthy tissue is also attacked by these particles, causing symptoms of muscle ache, joint ache, and headache. Therefore, everything aches so severely in patients infected with the Ebola virus. The best that can be done to assist patients in managing these symptoms is to offer the patient anti-inflammatories and painkillers. Unfortunately, even during this stage of the disease, it's still possible to write off the symptoms as being that of another form of infection, such as malaria. After being examined by a physician, it is at this stage that the presence of Ebola is discovered.

The minute the patient develops the onset of Ebola, they become weak and fatigued. The energy of the infected individual is severe, causing them to be confined to bed, unable to do anything but go through the experience of the virus's attack on their body. Fatigue and weakness continue to progress throughout the infection. These symptoms must be treated as soon as possible. If left untreated, the patient may develop seizures and fall into a coma from the combination of high fever and fatigue. At this stage of the disease, the patient will require 24-hour managed care (Zeng, 2021).

Perhaps some of the more dire symptoms of Ebola are diarrhea and vomiting. When the Ebola virus attacks the digestive system, hemorrhaging and internal bleeding occur in the intestines. This can cause the patient to start vomiting blood because of the infection in the upper gastrointestinal (GI) tract (also referred to as the gut) and stomach. In addition to this, the patient infected with Ebola may begin to experience bloody diarrhea (Zeng, 2021).

Greater than 70 percent of the immune system is in the gut. The intensity of these symptoms may differ depending on the type of strain responsible for the outbreak. The health of the "good bacteria" in the gut also contributes to the intensity of the symptoms. When Ebola becomes dominant in this region of the body, the patient literally begins the start of the fight for their life. At this stage of the Ebola virus invading the individual's body, any family members or friends who have encountered the patient over the course of the last few days, should be required to quarantine to ensure they do not have the virus (Zeng, 2021).

One of the first signs of severe hemorrhaging that occurs when a patient has been diagnosed with Ebola is when the whites of the eyes begin to turn red. The infected individual exhibits hemorrhaging of the blood vessels around the eyes, resulting in red eyes due to pooling of blood in the tear ducts. It appears that the patient is crying bloody tears when the patient blinks. This is characteristic of hemorrhagic fever. Chest pain and difficulty breathing occur during the final stages of Ebola virus infection. When Ebola spreads to the cardiovascular and respiratory systems, it causes the patient in the advanced stage to exhibit hemoptysis (coughing up blood) frequently in large volumes (Zeng, 2021).

Blood pools in the lungs as the tissues in the lungs begin to bleed out. This causes the patient to start to experience a tightness in the chest accompanied by chest pain as they begin to drown in their plasma. As the patient attempts to clear their chest, the result is they end up coughing up blood instead of the phlegm that would typically be expected when dealing with a bronchial or influenza infection. Anyone providing direct care to a patient at this stage of infection should do so by donning a hazmat suit. Because blood particles may vaporize into the air when the patient coughs, this could leave any caregiver exposed to contracting the infection by breathing in contaminated particles (Zeng, 2021).

During the final stage of the Ebola virus, the patient bleeds from all orifices. The consensus is that once the patient reaches this stage of the disease, there is minimal chance of surviving the level of internal hemorrhaging going on with the patient's organs. As the bleeding occurs throughout the body, the patient bleeds from the eyes, ears, mouth, and anus as the virus starts to manifest itself into a tragic form of death. The patient may die from choking on plasma collecting in the lungs or perish due to severe blood loss. The patient may slip into a coma, or seizures may start to occur prior to death. When the patient dies, the World Health Organization (WHO) mandates specific guidelines to be followed when disposing of the patient's body (Zeng, 2021).

Ebola remains one of the most intriguing infectious diseases on the planet. Its hemorrhagic symptoms are different from any other type of viral infection. This is the primary factor which makes it such a mysterious illness with horrific results. Approximately ten years have passed since the world's largest Ebola virus outbreak, originating in West Africa in late 2013. Medical Teams (an organization of medical professionals) was on the front lines then, and during a 2018 Ebola outbreak originating in the Democratic Republic of the Congo (DRC). Now, this organization is remaining vigilant and ready to respond in light of recent news of another potential outbreak (Zeng, 2021).

GREENHOUSE GASSES

Greenhouse gas emissions are atmospheric concentrations of carbon dioxide, methane, nitrous oxide, and synthetic fluorinated chlorofluorocarbons (Denchak, 2019). According to Denchak, (2019) "human activities such as burning fossil fuels and deforestation have caused a jump in the concentration of greenhouse gasses in the atmosphere in the last century. This has resulted in extra trapped heat and higher global temperatures."

The start of the Industrial Revolution saw the beginning of coal-powered steam engines, and human activities which greatly increased the volume of greenhouse gases released into the atmosphere. Between the years 1750 and 2011, it is estimated that concentrations of carbon dioxide in the atmosphere increased by 40 percent, methane increased by 150 percent, and nitrous oxide increased by 20 percent (Denchak, 2019).

There has been an increase in manufactured emissions of carbon dioxide. Of all the emissions, this is the greatest amount of greenhouse gas released by human activities, as well as one of the long-lasting, from 1750 to 2010. In the last 40 years, approximately half of these greenhouse gases alone were generated. This was due in large part to fossil fuel combustion and industrial processes. Between 2014 and 2016, global greenhouse gas emissions have infrequently plateaued from year to year. Currently, they are increasing again. Notably, in 2017, carbon emissions rose by 1.6 percent; in 2018, they increased by an estimated 2.7 percent (Denchak, 2019).

Greenhouse gases have long been debated by astrophysicists and climate scientists for decades. Different views and opinions have been expressed and deliberated to agree on or dispute the facts about greenhouse gas emissions. Figure 5.1 depicts how greenhouse gases warm our planet.

There are five major greenhouse gases that are most significant in causing global warming via the greenhouse effect. These are carbon dioxide, methane, nitrous oxide, fluorinated gases, and water vapor.

Carbon dioxide (CO_2) accounts for approximately 76 percent of global human-caused emissions. It remains for a period after it is released into the atmosphere. After it has been released into the atmosphere, 40 percent of the gas is still present after 100 years. Twenty percent is still present after 1,000 years, and 10 percent after 10,000 years (Denchak, 2019).

Methane (CH_4) remains in the atmosphere for approximately ten years. However, it is much more astringent in terms of the greenhouse effect. Pound for pound, its global warming impact is 25 times greater than that of carbon dioxide over a 100-year period. Globally, it accounts for about 16 percent of human-generated greenhouse gas emissions (Denchak, 2019).

Another potent greenhouse gas is nitrous oxide (N_2O). Its global-warming potential (GWP) is 300 times more potent than that of carbon dioxide on a

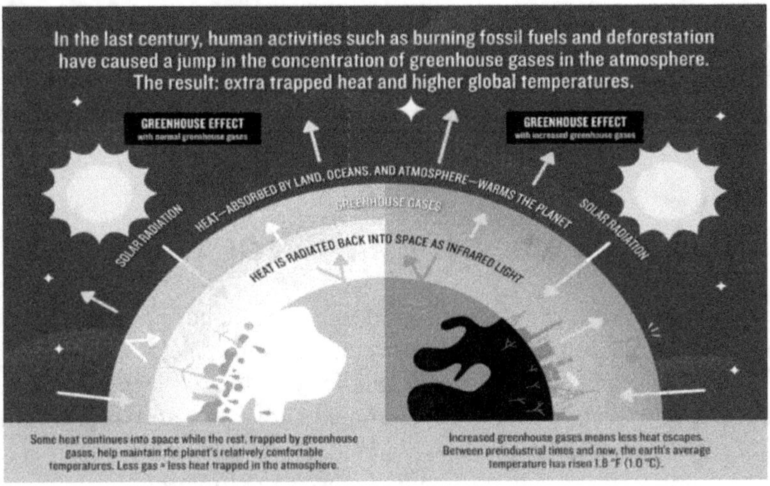

Figure 5.1 How human activity affects greenhouse gas emissions. Provided by the Environmental Protection Agency (EPA) and the Intergovernmental Panel on Climate Change [IPCC] (2014) (Denchak, 2019).

100-year time scale. Nitrous oxide remains in the atmosphere, on average, for slightly over a century. It accounts for nearly six percent of human-caused greenhouse gas emissions worldwide (Denchak, 2019).

Fluorinated gases are manufactured. They are produced from a variety of manufacturing and industrial processes. According to Denchak (2019), "There are four main categories: hydrofluorocarbons (HFCs), perfluorocarbons (PFCs), sulfur hexafluoride (SF_6), and nitrogen trifluoride (NF_3)". Fluorinated gases are emitted in significantly smaller quantities than other greenhouse gases, accounting for only two percent of fabricated global greenhouse gas emissions. Additionally, these gases trap considerably more heat (Denchak, 2019). The GWP for these gases can be in the thousands to tens of thousands. These gases have long atmospheric lifetimes, in some cases lasting tens of thousands of years. (Denchak, 2019).

In air conditioners and refrigerators, HFCs are used to replace ozone-depleting chlorofluorocarbons (CFCs) and hydrochlorofluorocarbons (HCFCs). Due to the high GWP, some of these are being phased out. One of the most crucial steps the world can take is to replace these HFCs and dispose of them properly (Denchak, 2019).

Melissa Denchak (2019) posits, "Water vapor is the most abundant greenhouse gas overall. It differs from other greenhouse gases in that changes in its atmospheric concentrations are linked not to human activities directly, but rather to the warming that *results* from the other greenhouse gases we emit." The warmer the is, the more water it holds. Water vapor is one of the greenhouse gases; more water holds more heat, producing substantially greater warming and maintaining a positive feedback loop. Notably, the net effect of this feedback loop remains uncertain. The rationale which ensues, an increase in water vapor fosters an increase in cloud cover that reflects the sun's energy away from the earth (Denchak, 2019).

THE DIGITAL TRANSFORMATION OF HEALTHCARE

There have been few technological developments that have impacted communication and the dissemination of information as the internet and the World Wide Web. This galvanizing new technology is having a grave impact on the health care industry. Patient education and advocacy services, continuing medical education, telemedicine, and digital agents and telediagnosis are only a minute number of areas that involve the use of the internet.

Whether there is an agreement or not, it is keenly recognized that health care has become a high transaction business. Approximately 15 percent of the entire data transmissions in the United States are associated with health care (Colie, 2000). The number of transactions that can range from a routine physician visit or a complex cardiac work-up can vary from five-to-twenty-three. With the advent of the internet, it has been proven that the cost of transactions can be lowered from $5 per patient to $60 per patient. This could yield an annual savings of $10,000 for a physician who sees 500 patients monthly (Colie, 2000).

The internet has become the frontier of the health care industry. It's been reported by the Harris Poll that 74 percent (70 million) of the estimated 91 million people in the last year visited at least one of the 20,000 health-related sites on the Web seeking medical information (Colie, 2000). Commonly referred to as "e-health," the digital transformation of health care has profoundly reshaped not only health care but banking, airlines, and retail shopping.

Some consumers of health care are ardently using the Web searching globally for the most up-to-date medical research or treatment data that has been evaluated. The current health care market offers access to drug refills, durable medical equipment (DME), and alternative medicines. According to Colie, (2000), "an estimated 60 percent of consumers using health websites have purchased other products online" (p. 10). Well-known health care providers such as MD Anderson hospital in Houston, TX, and the Cleveland Clinic in Cleveland, OH, acknowledge that the baby boomer generation readily turns to these facilities for digital information. This is attributed to the well-developed internet websites available and user-friendly to a generation that has become the quintessential loyal consumers of digital medical information.

According to the Health Information Management Systems Society (2000), hospitals and other health care systems are beginning to direct the resources they offer on Web-enabled, e-commerce applications to different areas in health care, such as marketing, physician directories, employee recruitment, and buying supplies online by comparison shopping on the internet via companies like medicalbuyer.com. Only six percent of the health care industry utilizes the internet to purchase supplies online today. Furthermore, it has been estimated that 25 percent of America's $1.2 million health budget is deemed to be extremely excessive administrative costs or unnecessary medical treatments. The digital marketplace offers medical products via the internet, which can reduce costs by five to fifteen percent or more when purchased online (Colie, 2000).

What is needed in the health care industry is a low-cost technology that connects competing hardware and software, providing a "seamless web" relative to communication pathways. New information resources and transactions for payers, purchasers, health care providers, and suppliers are being created by Dot.com companies. The goal is to have as many of these Dot.com companies as possible to accommodate the need of the health care industry.

The internet and the World Wide Web have both been identified as extremely cost-productive in health care. Conversely, the greatest barrier to the pervasive adoption of e-health strategies as a common occurrence may be health care's unstable era of the "Balanced Budget Act (BBA)." Medicare's cutback of payments under the BBA has caused hospitals and physicians to be reluctant to invest in internet strategies.

The question then becomes, how will the internet and the World Wide Web be of greater use in the future? The new e-health economy is moving into a second generation, going beyond the "Business-to-customer" (B2C). The "Business-to-Business" (B2B) strategies base revenues on transactions or subscriptions. The belief is that web-based solutions may apply to both sides, in essence, providing a "balance sheet" to decrease expenditures, and creating new sources of revenue. There is an expectation that e-health strategies will increase exponentially within the next five years. America's healthcare executives will be compelled to change the method of practice using information systems/information technology as a fundamental business and clinical process in operating the health care industry.

CONCLUSION

The twenty-first century has brought about many changes regarding perspectives on climate change, pandemic(s), and digital transformation, particularly where health care is concerned. As American culture changes, sentiments such as “the new normal and “in these uncertain times,” are spoken frequently. There seems to be public discourse, exhibiting a general feeling of vulnerability and transformation on a national level.

Multiple catalysts are usually the reason for change. The specific aspects which have been addressed in this chapter focus on some of the issues associated with shifts that have contributed to changes manifested in the twenty-first century. Another perspective that offers an explanation is correlated with a form of Adorno’s (2015) administered world. The belief here is that we place society in order according to dominant opinions. The general meaning is those well accustomed to institutional and corporate culture have the best opinion of what is best for the “good of the masses.” However, Hannah Arendt (1963) in her book, “The Promise of Politics”, was concerned that matters deemed to be personal, were becoming political. Arendt further explained that “the toxic public discourse invades private discourse, reducing personal autonomy and leaving us perpetually under the scrutiny of institutional authority” (Arendt, 1963). Whether we agree or disagree with Adorno or Arendt, the general premise remains; the last ten years of the twentieth century profoundly impacted how we live in the twenty-first century.

REFERENCES

Bernow, S, Duckworth, M. (1998). An Evaluation of Integrated Climate Protection Policies for the United States. *Conservation Ecology*, 2(2).

Coile Jr., R. C. (2000). The Digital Transformation of Health Care. *Physician Executive*, 26(1), 8.

Cook, J. (2016). Ten Years on: How Al Gore’s An Inconvenient Truth Made Its Mark, *Global Change Institute,* The University of Queensland

Denchak, M. (2019). Greenhouse Effect 101

Doyle, D. J., Ruskin, K. J., Engel, T. P. (1996). The Internet and Medicine: Past, Present, and future. Yale, J Biol Med. Sep-Oct;69(5): 429-37. PMD: 9381738. PMCID: PMC2589000.

Simmons. J., Nordenhaug, E., McIntyre, K. B., Rich, L. E., Butterfield, E., Ulbig, S. G. & Gressis, R. (2020). The Twenty-First Century and Its Discontents: How Changing Discourse Norms are Changing Culture

Singer, F., Legates, D., Lupo, A. (2021). Hot Talk, Cold Science: Global Warming’s Unfinished Debate. Barnes & Noble, New York, New York.

Sullivan, L. (2021). What is Ebola & How Does Ebola Spread?

Tickell, C. (1990). Huan Effects of Climate Change: Excerpts from a Lecture Given to the Society on 26 March 1990. *The Geographical Journal*, 156(3), 325-329 https://doi.org/10.2307/635534

Weiss, M. (1997). Signifying the Pandemics: Metaphors of AIDS, Cancer, and heart disease. *Medical Anthropology Quarterly*, 11(4), 456-476. https: www.jstor.org/stable/649534

Zeng, K. (2021). Study reveals Ebola virus can hide in brain, persist even after treatment. *U. S. Army Medical Research Institute of Infectious Diseases*.

Section 2

THE TWENTY-FIRST CENTURY

A PRESENT AND FUTURISTIC VIEW

Chapter 6

Education in the 21st Century

The Quest for Sustainable Education in the Global South

By Kofo A. Aderogba & Tolu O. Isele

INTRODUCTION

Education is a social institution through which children of a society are taught basic academic knowledge, learning skills, and cultural norms. Every nation is equipped with some form of education system, though those systems vary greatly. Significant factors that affect education systems are the resources and money utilized to support those systems in the different nations, as well as how it is systematized and managed (Lumenwaymaker, n.d.). Expectedly, the wealth of a country has much to do with the amount of money spent on education. Countries that do not have such basic amenities as running water cannot support robust education systems or, in many cases, any formal schooling (OpenStax, 2022). OpenStax (2022) asserts further that the consequences of this worldwide educational inequality are of social concern for many countries, including the United States and other developed worlds.

International differences in education systems may not solely be due to any financial issue. The value placed on education, the amount of time devoted to it, and the distribution of education within a country also play a role in the differences. For instance, students in South Korea spend an average of 220 days a year in school, compared to the 180 days a year of their United States' counterparts (Pellissier 2013). As of 2006, the United States ranked fifth among twenty-seven countries for college participation but ranked sixteenth in the number of students who receive college degrees (National Center for Public Policy and Higher Education, 2006). These statistics may be related to how much time and resources are spent on education.

Results from the most recent Programs for International Student Assessment (PISA) in 2015, which is a test administered to 15-year-old students in participating economies every three years, showed that students in the United States lag far-off behind other high-income countries across different subjects, despite making the most progress in equity compared to other participating countries (PISA, 2015). Countries in the Asian and Nordic regions tend to outperform other countries across all subjects assessed (Science, Mathematics, and Literacy), although their educational systems are immensely different in their goals and their administrative systems, suggesting that there is not a single recipe for "success" as measured by PISA (Lumenwaymaker, n.d.).

Analysts determined that the nations and city-states at the top of the rankings had several things in common. They had well-established standards for education with clear goals for all students. Although these countries have well-delineated standards, they do not necessarily outline similar goals (Lumenwaymaker, n.d.). For example, one country may emphasize cooperation, another student growth, or yet another may focus on equality, as in Finland (Gross-Loh, 2014). Another thing the high-performing nations had in common was a tendency to recruit teachers from the top 5 to 10% of university graduates each year, which is not the case for most countries (National Public Radio, 2010).

There is also the issue of social factors: One analyst from the Organization for Economic Cooperation and Development attributed 20% of performance differences and the low rankings of the United States to differences in social background. Researchers noted that educational resources, including money and quality teachers, are not distributed equitably in the United States. In the top-ranking countries, limited access to resources did not necessarily predict low performance. Analysts also noted what they described as "resilient students," or those students who achieve at a higher level than one might expect given their social background. In Shanghai and Singapore, the proportion of resilient students is about 70%. In the United States, it is below 30%. These insights suggest that the United States' educational system may be on a descending trajectory that could detrimentally affect the economy and social landscape of the country (National Public Radio, 2010).

Recent research has found that the low overall educational achievement in United States (US) is largely due to underperformance by the Middle Class. The poorest students in the United States, despite being among the most socioeconomically disadvantaged around the world, perform averagely relative to other poor students. The most affluent students from the US, despite being among the wealthiest, are also average when compared to other rich students - which is also an alarming finding. But students in the middle of the Socioeconomic Status (SES) distribution perform half a school year behind comparable middle-SES students, despite being among the wealthiest middle-SES groups in the world (Rowley et al., 2019; Lumenwaymaker, n.d.).

The low level of education in the Global South has primarily been attributed to poor policy-making (Kayagil et al., 2018). African nations, Latin America, Mexico, and others are significantly behind in education standards and performance levels (Kayagil et al., 2018). They have the most prominent teachers' unions, there are government ministries and agents of education saddled with the responsibility of ensuring sustainable education at all levels of learning, and with deep historical and political roots. The players in the sector have not been able to pursue the interest of the other stakeholders. They have consistently fell behind initiatives in education reforms and others. There is thus a wide gap in the standards between the Global South and the Global North (Kayagil et al., 2018; Lumenwaymaker, n.d.; Ukaigwe & Chima, 2019). Over the past decades, the sector has been characterized by conflicts and strikes. Teachers spend months without seeing the doors of any learning center. Students remain at homes fraternizing with their parents and guardians. There was lack of infrastructure at some instances, and horrendous decay at the others. The budgetary allocations and expenditures are often short of the 26% myth prescriptions of the United Nations Educational, Scientific and Cultural Organization (UNESCO) (Ukaigwe & Chima, 2019). The recent pandemics of COVID-19 compounded the already devastating conditions with the over 1.6 billion learners locked out from schools around the world.

The twelfth edition of the Education for All (EFA) Global Monitoring Report - marking the 2015 deadline for the six goals set at the World Education Forum in Dakar, Senegal, in 2000 - provides a considered and comprehensive accounting of global progress in education. As the international community prepares for a new development and education agenda, the report takes stock of past achievements and reflects on future challenges. There are many signs of notable advances. The pace towards universal primary education was quickened, gender disparity was reduced in many countries and governments increased their focus on ensuring children receive quality education. However, despite the efforts and the so-called achievements, the world failed to meet its overall commitment to Education for All: Millions of children and adolescents still remain out of school, and it is the poorest and most disadvantaged who bear the brunt of the failure to reach the EFA targets. The level of achievement in the Global South is less than 15% of the total (UNESCO, 2015, 2018). Though the UNESCO (2015) Education for All 2000-2015 assessment provides a comprehensive assessment by countries, progress towards the EFA goals and highlights of the works that remains for attaining sustainable education, there is still much, and a long way to go in the Global South.

However, the chapter has looked into the history of education with particular emphasis on the Global South, the challenges, and opportunities; and it looks into the future for development of education in the Global South in the 21st Century. It is in six parts. Data and information were drawn from literature, records of governments and government agencies, and Departments and Commissions of the United Nations. It is devoid of any radical inclinations; but some histories were traced to elucidate salient background to the contemporary issues.

The Global South

The term 'Global South' emerged in the 1950s, but Carl Oglesby became the first person to give it a contemporary political use when he commented on the dominance of US over the Global South (Otieno, 2019). The founding members of the Non-Aligned Movement (NAM) also used the term politically (Otieno, 2019; Hollington et al., 2021). To better understand the term, one needs to know the North-South divide's social, economic, and political meaning. The North-South divide does not mean a division along the equator, but the imaginary line dividing the richest and the poorest countries on the planet Earth. Global North, therefore, are the developed countries of North America, Europe, Asia, and Oceania (The West, the First World, and parts of the Second World). The Global South, therefore, includes countries in Africa, Latin America, and developing parts of Asia and the Middle East. Alternative terms for the Global South are the Less-Developed World, Developing Countries, Majority World, Non-Western World, Poor World, the South, Third World, and the Undeveloped World (Otieno, 2019); also see Figure 6.1.

It is a dynamic term that does not consider geographic locations, meaning that members of the group who reach a certain development threshold

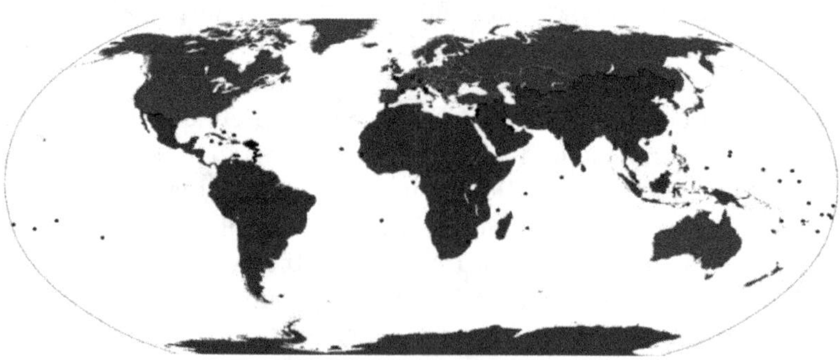

Figure 6.1 World map showing a traditional definition of the North–South divide (Red countries in the map are grouped as "Global South"; Blue countries are "Global North".) Hollington, A., Tappe, O., Salverda, T. & Schwarz, T. (2021 November 21). Introduction: Concept of the Global South. Global South Study Center Cologne.

may cross over to the Global North. Even in the Global North, some regions within the developed countries live in conditions resembling the Global South's conditions (Otieno, 2019). Ironically, Australia is in the Global North despite the substantial Global South between it and the North.

After the initial references for the term, it has emerged as a unifying identity that politically and economically brings together the countries within the Southern Hemisphere. The accord also extends to cultural, social, technical, and environmental cooperation in the framework called South-South Cooperation (SSC). The foremost objective of SSC is to pursue mutually beneficial economic changes that unite the Global South within the manipulative world system. The SSC respects values such as non-interference in domestic matters, non-conditionality, independence, sovereignty, equality, and national ownership. Formerly, SSC proposed to use the cooperation to address challenges like poverty, cross-border issues, population growth, disease, and war through sharing skills, experiences, resources, and expertise.

One of the most noticeable results of SSC is the determination of countries like China and India to take over the economic and political dominance of the west in most Global South countries. Indeed, China's trade and development cooperation with most African countries has surpassed the cooperation with the same countries in the West. This situation continues to give China a bigger voice and influence in the affairs of such countries.

Actors in international relations favour the use of the term "Global South" as compared to other terms like "Least Developed" or "Third World" or "Developing Countries." Indeed, most scholars believe that the term, not only resists backwardness tendencies of global dominance but also encourages a rethink of the relationship between the Global North and Global South from cultural and developmental difference to geopolitical benefits and relationships (Mahler, 2018; Milian, 2013). This school of thought aims to correct the negative impact of contemporary global capitalism as well as colonial and neo-imperial histories that led to some level of poverty and inequality in the Global South (Lees, 2020; Mahler, 2018; Milian, 2013; Otieno, 2019). Conversely, critics argue that the term Global South does not benefit all countries in the defined bracket but only the rich ones within the group. In their critique, scholars believe that the developed countries within the Global South politically and economically exploit their underdeveloped counterparts within SSC international relations. There is a third school of thought that critiques the term because the majority of the Global South countries are actually located in the north of the equator, therefore, they do not feel attached to the term. Also, the grouping of the Global South has no cultural, historic, or economic significance between the regions, but developmental imbalances (Otieno, 2019).

West-Pavlov & Connell (2018) summarized it all this way:

"…. the regions of Latin America, Asia, Africa, and Oceania. …. family of terms, including "Third World" and "Periphery," that denote regions outside Europe and North America, mostly low-income and often politically or culturally marginalized. …. Global South marks a shift from a central focus on development or cultural difference toward an emphasis on geopolitical relations of power. …. social progress by drawing broad distinctions between "advanced" and "primitive" institutions and societies. …. The societies identified in early sociological texts as the most primitive were the San people of Southern Africa, the indigenous people of Tierra del Fuego, and the Aboriginal people of Australia. It is no accident that these were three of the southernmost populations known to European colonizers; literally the most distant. …. "The Southern Question" began with the idea that southern Italy had, in effect, been colonized by capitalists from northern Italy. Gramsci explored the difficulties southern peasants and northern workers faced in forging an alliance with one another. In Italian social thought, most later treatments of the "North South differential" addressed only variations in economic development. …. Raúl Prebisch popularized the distinction between "core" and "periphery" of the world economy. …. analyzed underdevelopment and struggled for reform of the world trade system. This struggle moved the terms "North" and "South" into the international political lexicon. …. Developing countries began to articulate the idea of a Global South whose interests conflicted with those of the industrialized powers, both capitalist and communist - cutting across Cold War divisions. …. a "world-system approach" that made the concepts "core" and "periphery" alternatives to the "modern/traditional" binary. …. in Marxist economics.

According to West-Pavlov & Connell (2018), these concepts were reinforced by "intersectional" perspectives in northern sociology, especially those of African American and Chicana/o scholars. They inferred:

"…. With the Cold War winding down, the terms …. spread in academic fields like international relations, political science, and development studies. The North South language provided an alternative to the concept of "globalization," contesting the belief in a growing homogenization of cultures and societies. A powerful Global North and a resistant Global South was promoted …. "Global South" marks a shift from a focus on development or cultural difference toward an emphasis on geopolitical power relations. Industrial growth in Asia, and the emerging Brazil, Russia, India, China (BRIC) group complicated the North-South picture. But the huge growth of finance capital centered in New York, London, Frankfurt, and Tokyo reinforced the economic advantage of the old imperial centers. …. the North-South concept has acquired increased sophistication. …. North-South terminology, then, like core-periphery, arose from an allegorical application of categories to name patterns of wealth, privilege, and development across broad regions. Global South functions as more than a metaphor for underdevelopment. It references an entire history of colonialism, neo-imperialism, and differential economic and social change through which large inequalities in living standards, life expectancy, and access to resources are maintained."

These views are not so significantly different from those of Mahler (2018), Milian (2013) and Prashad (2012), in parts, and of similar communities.

However, it is of relevance to reiterate and highlight that there are different opinions with regard to various aspects of the term and that it evokes different meanings for different people; meanings, moreover, that may shift over time. The Global South is contextual, as most contributions highlight. In times of geopolitical uncertainty, it is hard to predict how the term will develop and/or change accordingly. The question is whether it will actually become an obstacle to a more equal distribution of the world's gains and power or whether it might actually empower parts of the world that have a long history of disadvantage. It would imply that the mere use of the term might have implications, for better or for worse. But the increasing usage of the concept might also reflect changing realities: the terms that seem convenient to describe the reality of specific historical moments are closely related to the respective socioeconomic and political structures (Dados & Connell, 2012; Hollington et al., 2021; United Nations Finance Center for South-South Cooperation, 2022).

The History of Education Around the World

The history of education extends at least as far back as the first written records recovered from ancient civilizations. Historical studies have included virtually every nation (Bowen, 1981; McCulloch & Crook, 2013; Baker et al., 2010). Possibly the earliest formal school was developed in the Middle Kingdom of Egypt under the direction of Kheti, treasurer to Mentuhotep II (2061-2010 BC) in the Middle East (Parsons, 2022). In Mesopotamia, the early logographic system of cuneiform script took many years to master. Accordingly, only limited number of individuals were employed as scribes to be trained in its reading and writing. Only royal offspring and sons of the rich and professionals were schooled (Thomason, 2005). Most boys were taught their fathers' trades or were apprenticed to learn a trade (Harris, 2003). Girls stayed at home with their mothers to learn housekeeping and cooking, and to look after the younger children. Later, when a syllabic script became more widespread, more of the Mesopotamian population became literate. Later still in Babylonian times, there were libraries in most towns and temples; probably the source of the old Sumerian proverb that avers that "he who would excel in the school of the scribes must rise with the dawn." There arose a whole social class of scribes, mostly employed in agriculture, but some were employed as personal secretaries or solicitors (Fischer, 2004). Women as well as men learned to read and write, and for the Semitic Babylonians, this involved knowledge of the extinct Sumerian language, and a complicated and extensive syllabary. Vocabularies, grammars, and interlinear translations were compiled for the use of students, as well as commentaries on the older texts and explanations of obscure words and phrases. Massive archives of texts were recovered from the archaeological contexts of Old Babylonian scribal schools known as edubas (2000–1600 BCE), through which literacy was disseminated (McIntosh, 2021). The Epic of Gilgamesh, an epic poem from Ancient Mesopotamia is among the earliest known works of literary fiction. The earliest Sumerian versions of the epic date from as early as the Third Dynasty of Ur (2150–2000 BC) (Benavot & Julia, 2006; McIntosh, 2021; Oloruntimehin, 1974).

Ashurbanipal (685 – c. 627 BC), a king of the Neo-Assyrian Empire, was proud of his scribal education. His youthful scholarly pursuits included oil divination, mathematics, reading and writing as well as the usual horsemanship, hunting, chariotry, soldierliness, craftsmanship, and royal decorum. During his reign, he collected cuneiform texts from all over Mesopotamia, and especially Babylonia in the library in Nineveh. The first systematically organized library in the ancient Middle East (Baines, 1983). Some of these survive till date.

Similarly, in ancient Egypt, literacy was concentrated among an educated elite of scribes. Only people from certain backgrounds were allowed to train to become scribes, in the service of temple, pharaonic, and military authorities. The rate of literacy in Pharaonic Egypt during most periods from the third to first millennium BC was estimated at not more than one percent, or between one half of one percent and one percent (Baines, 1983; Hopkins, 1991). In ancient Israel, the Torah (the fundamental religious text) includes orders to read, learn, teach, and write the Torah, thus requiring literacy and study. In 64 AD the high priest caused schools to be opened (Hopkins,1991). Emphasis was placed on developing good memory skills in addition to comprehension and oral repetition (Compayre & Payne, (2003).

In the Islamic civilization that spread all the way between China and Spain during the time between the 7th and 19th centuries, Muslims started schooling from 622 in Medina, which is now a major city in Saudi Arabia. Schooling at first was in the mosques (masjid in Arabic) but then schools became separate and next to mosques. The first was the Nizamiyah school built in 1066 in Baghdad where children started school from the age of six, tuition free. The Quran encourages Muslims to be educated. Thus, education and schooling sprang up in the ancient Muslim societies. Moreover, Muslims had one of the first universities in history which is Al-Qarawiyin University in Fez, Morocco. It was originally a mosque that was built in 859 (Al-Hassani, 2011).

In the Indian Subcontinent, the development were closely similar: Education was mainly imparted through the Vedic and Buddhist education system. Sanskrit was the language used to impart the Vedic education system. Pali was the language used in the Buddhist education system. In the Vedic system, a child started his education at the age of five, whereas in the Buddhist system, the child started his education at the age of eight. The main aim of education in the ancient India was to develop a person's character, master the art of self-control, bring about social awareness, and to conserve and take forward ancient culture. The Buddhist and Vedic systems had different subjects: In the Vedic system of study, the students were taught the four Vedas (Rig Veda, Sama Veda, Yajur Veda and Atharva Veda). They were also taught the six Vedangas (ritualistic knowledge, metrics, exegetics, grammar, phonetics, and astronomy), the Upanishads and more. In the Buddhist education system, the subjects included Pitakas.

In China, the rulers, Yao and Shun (ca. 24th-23rd Century BC) established the first schools. The first education system was created in Xia dynasty (2076–1600 BC). During his dynasty, government-built schools to educate aristocrats about rituals, literature, and archery (important for ancient Chinese aristocrats). During Shang dynasty (1600 BC to 1046 BC), normal people (farmers, workers, and others) accepted "rough" education. The children of aristocrats studied in government schools. And normal people studied in private schools. Government schools were built in cities and private schools were built in rural areas. Government schools paid attention to educating students about rituals, literature, politics, music, arts, and archery. Private schools educated students to do farm work and handworks (Lin, 1993).

In the subsequent centuries and with the subsequent rulers, there were massive developments: After learning Chinese characters, students wishing to ascend in the social hierarchy needed to study the Chinese classic texts. The early Chinese state depended upon literate, educated officials for operation of the empire. In 605 AD, an examination system was explicitly instituted for a category of local talents. The merit-based imperial examination system for evaluating and selecting officials gave rise to schools that taught the Chinese classic texts and continued in use for about 1,300 years. It was only abolished in 1911 in favour of Western education methods. The core of the curriculum for the imperial civil service examinations from the mid-12th Century onwards was the Four Books, representing a foundational introduction to Confucianism. Hypothetically, any male adult in China, regardless of his wealth or social status could become a high-ranking government official by passing the imperial examination, although under some dynasties, members of the merchant class were excluded. Nevertheless, most of the candidates came from the numerically small but relatively wealthy land-owning elites (Lin, 1993).

In the period preceding 1040-1050 AD, prefectural schools had been neglected by the state and left to the devices of wealthy patrons who provided private finances (Yuan, 1994). Girls were educated in the crafts of home and child raising. They were not taught to read or write. All women were taught to be involved in religion; there were paintings of women presiding over religious ceremonies, but there were no references to female priests. In Central and South American Civilization, Aztec teachers propounded a spartan regime of education with the purpose of forming a stoical people (Kagan et al., 2007).

Until the age of fourteen, the education of children was in the hands of their parents but supervised by the authorities of their calpolli. Part of this education involved learning a collection of "sayings the old" that embodied the Aztecs' ideals. Most of these seemed to have evolved over several centuries, predating the Aztecs and most likely adopted from other Nahua cultures. At 15, both boys and girls went to school. The Mexica, one of the Aztec groups, were one of the first people in the world to have mandatory education for nearly all children regardless of gender, rank, or station. There were two types of schools: the telpochcalli, for practical and military studies, and the calmecac, for advanced learning in writing, astronomy, statesmanship, theology, and other areas. The two institutions seem to be common to the Nahua people, leading some experts to suggest that they are older than the Aztec culture.

After the 15th Century, there were great developments in China, Europe; and by the late 19th Century, most of West, Central, and parts of East Europe began to provide elementary education in reading, writing, and arithmetic, partly because politicians believed that education was needed for orderly political behavior. As more people became literate, they realized that most secondary education was only open to those who could afford it. Having created primary education, the major nations had to give further attention to secondary education by the time of World War I (Kagan et al., 2007).

In the 20th Century, new directions in education included, in Italy, Maria Montessori's Montessori schools, and in Germany, Rudolf Steiner's development of Waldorf education. In Canada, education became a contentious issue after Confederation in 1867, especially regarding the status of French schools outside Quebec. Education in New Zealand began with provision made by the provincial government, the missionary Christian churches and private education. The first act of parliament for education was passed in 1877 and sought to establish a standard for primary education. It was compulsory for children to attend school from the age of 6 until the age of 16 years (Education.govt.nz, 2020; Paglayan, 2021).

In Australia, compulsory education was enacted in the 1870s, and it was difficult to enforce. People found it hard to afford school fees. Moreover, teachers felt that they did not get a high salary for what they did (McCreadie, 2012).

In Imperial Russia, according to the 1897 census, literate people made up 28 percent of the population. There was a strong network of universities for the upper class, but weaker provisions for everyone else. In 1919, Vladimir Lenin proclaimed the major aim of the Soviet government - the abolition of illiteracy. A system of universal compulsory education was established. Millions of illiterate adults were enrolled in special literacy schools. Youth groups were utilized to teach. In 1926, the literacy rate was 56.6 percent of the population. By 1937, according to census data, the literacy rate was 86% for men and 65% for women, making a total literacy rate of 75% (Paglayan, 2021).

The fastest expansion of primary schooling in the history of the Soviet Union coincided with the First Five-Year Plan. The motivation behind this rapid expansion of primary education can largely be attributed to Stalin's interest in ensuring that everyone would have the skills and predisposition necessary to contribute to the state's industrialization and international supremacy goals. Indeed, Paglayan (2021) notes that one of the things that most surprised US officials during their education missions to the USSR was, in their words, "the extent to which the Nation is committed to education as a means of national advancement. In the organization of a planned society in the Soviet Union, education is regarded as one of the chief resources and techniques for achieving social, economic, cultural, and scientific objectives in national interest. Tremendous responsibilities are therefore placed on Soviet schools, and comprehensive support is provided for them" (Education .govt.nz, 2020; Paglayan, 2021).

An important aspect of the early campaign for literacy and education was the policy of "indigenization." This policy lasted from the mid-1920s to the late 1930s, promoted the development and use of non-Russian languages in the government, the media, and education. Intended to counter the historical practices of Russification, it had as another practical goal assuring native-language education as the quickest way to increase educational levels of future generations. A huge network of so-called "national schools" was established by the 1930s, and this network continued to grow in enrollments throughout the Soviet era. Language policy changed over time, perhaps marked first of all in the government's mandating in 1938 the teaching of Russian as a required *subject* of study in every non-Russian school, and then especially beginning in the latter 1950s a growing conversion of non-Russian schools to Russian as the main medium of instruction.

Education in French controlled West Africa during the late 1800s and early 1900s was different from the nationally uniform compulsory education of France in the 1880s. "Adapted education" was organized in 1903 and used the French curriculum as a basis, replacing information relevant to France with "comparable information drawn from the African context". For example, French lessons of morality were coupled with many references to African history and local folklores. The French language was also taught as an integral part of adapted education.

Africa has more than 40 million children. According to UNESCO's Regional overview on sub-Saharan Africa, in 2000 only 58% of children were enrolled in primary schools, the lowest enrollment rate of any region, that is, despite long history of development around the world. The United States Agency for International Development (USAID) Center reports as of 2005, 40% of school-aged children in Africa do not attend primary school. But by the end of the second decade of this Century, the situations have drastically improved in more than 55% of the states (UNESCO, 2015, 2017a, 2018; United Nations Children's Fund (UNICEF), 2020, 2021).

However, the history of education in the Global South takes its route from and intertwined with the history of the world's educational development. There have been humongous challenges.

Contemporary Issues in the Global South

UNICEF (2019) announced a "global learning crisis." UNESCO (2017b) also sounded the alarm: "More than one-half of children and adolescents are not learning worldwide." Yet, five years earlier, the United Nations claimed unprecedented success for its Millennium Development Goals, which included Education, because more children than ever before were in school. Advocates such as Bill Gates used this as clear evidence that development aid was working (Gate, 2018).

The quick answer is that in 2015, the United Nations needed to show great success for its original development goals (2000-2015) because it was seeking approval for another fifteen years of UN-led development policy. The UN "Education" goal measured only school enrollment, not whether children learned anything. There were loud voices protesting that school quality was getting worse, but the UN ignored them (Alyson, 2022; Arong & Ogbadu, 2010; Creffier, 2007). The protests might be right. While the UN promoted enrollment, overwhelming evidence shows a frightening drop in the quality of schools (Alyson, 2022; McGovern Dole International Food for Education and Child Nutrition Program, 2015; World Bank, 2013; Arong & Ogbadu, 2010):

> Kenya, 2008: The proportion of semi-literate or illiterate women after 6 years of schooling has worsened in recent years: In 2003, 24% were in this situation, compared with 39% in 2008." (Food for Education report, U.S. Department

> of Agriculture). The question is – almost 20 years after, what have been the remarkable differences?
>
> Nigeria, 2010: Education quality "from the primary up to the tertiary levels has significantly fallen. The products of primary schools are unable to write their own names." By 2021, there are mirage of challenges lowering the quality of education.
>
> Madagascar, 2013: Three assessments were done for grade 5 students, in 1998, 2005, and 2012. For French, scores fell from 42.6 to 31.4, and to 26.8, respectively. In Mathematics, they fell from 59.1 to 51.3, and to 40.0. (World Bank 2014). It has not been reported that there are remarkable improvement, at least, till the end of 2020.

However, these might be the situation around the world because of the followings too (Alyson, 2022; Economist, 2012, 2017; Time Magazine, 2014; Abadzi, 2017; Global Partnership for Education (GPE), 2021; UNESCO, 2021; Iyengar, 2014; New York Times, 2014; World Bank, 2017):

- India, 2014: More children are going to school in India, but they're learning less (Time Magazine, 2014).
- Thailand, 2012: "Although the proportion of children attending school has grown over the past decade, the quality of their education has deteriorated." (The Economist, 2017).
- Globally, 2015 of the UN development goals: "If the challenge was to provide a minimum standard of education for all, what looks like an enormous improvement, too often, amounted to a stunning failure." (New York Times, 2014).
- Globally, 2017: "The Education for All initiative greatly increased enrolments but gave rise, in some countries, to a generation of schooled illiterates." (World Bank Education Specialist, 2017).
- Globally, 2021: UNESCO reports: "Almost half of all 20- to 24-year-olds who had completed lower secondary school in 18 low- and lower-middle-income countries could not read a simple sentence."
- Globally, 2021: In 32 lower-income countries where the Global Partnership for Education works, "three out of every five students cannot read by the end of primary school."

These reports reveal only a piece of the crisis (Alyson, 2022). They focus on literacy, and sometimes Mathematics, as the standard measures of education quality. Yet even in these subjects, which receive the greatest effort, actual learning plummeted. What about history, geography, health, and science? No one reports how many high school graduates are unable to find any country except their own on a world map.

An even bigger concern is that children also need to develop critical thinking, problem-solving, and communication skills. They need to work well with others, and also take initiative and act independently when appropriate. They should learn to manage their emotions, and their time. Will children learn those things by sitting in a rote-based classroom, bored to death, as the teacher tells them what to remember for the test, then assigns two hours of homework as a false solution to the problem that they learn nothing in school (Abadzi, 2017; Alyson, 2022; Iyengar, 2014; GPE, 2021; Time Magazine, 2014; UNESCO, 2021).

In the 144 countries where UNICEF works, more than 50% of which are in the Global South, every report depicts mind-deadening, rote public schooling (Iyengar, 2014). This is the precise opposite of what children need if they are to thrive. For many children, it is psychological torture. Others tolerate it, and learn a bit, but could positively flourish in a setting that encouraged activity, curiosity, passion, experimentation, peer-help, and much else (Alyson, 2022). Even as it fails to develop their abilities, rote schooling makes hundreds of millions of children miserable, with no benefit to them. Enrollment numbers ignore this. The joys of childhood become collateral damage in the focus on filling the schools of the United Nations (UN).

The UN agencies and international NGOs are finally admitting the dismal state of public schooling in the countries where they pushed their policies for thirty years. But they carefully avoid any genuine discussion of why these schools have gotten worse. There is a good reason for that: When it announces a "global learning crisis," the UN can call urgently for money to combat the problem, as if this were another unpredictable calamity. But once we address the question of why schools have gotten worse, the evidence points right back to the UN and the aid industry as the major cause of this decline (Economist, 2017).

The dynamics of how aid funds and policies have momentously hurt the schemes in several ways abound. The following two are good examples of resultant school quality reports (Alyson, 2022):

- Donor priority: UN agencies and International Non-governmental Organizations (INGOs) are funded by donors, and their top priority is to satisfy those donors. It is easy to build schools and push children into them. It is not so easy to ensure that students learn something. Local officials can see that this is not working, but the aid industry offers personal enticements – ranging from phones to consulting fees to "fact-finding" travel – to follow priorities set from abroad and others.
- Churn: Donors also want something new. The constant churn of new, short-lived programmes leaves no time for local teachers and officials, who

understand of local conditions, to try out their own ideas. The aid industry has its own ideas, and the money to push through its priorities. In 2019, announcing the "crisis", UNICEF also put forward its Education Strategy for 2019-2030 (UNESCO, 2019). Scattered throughout these 64 pages are 57 "priorities" which UNICEF intends to push, including more data collection, more UNICEF education staff, and more tech-based classroom innovations.

Incidentally, directly, and otherwise, school quality in the Global South generally is being controlled from afar. No doubt those calling the shots do want better lives for the poorest children of the world, but at the same time, they probably do not want their own children to compete on a level field against a generation of those who are equally smart and work twice as hard. These mixed motivations arrive under the name of "aid," but the big decisions and priorities are made by the former colonizers, with disastrous consequences for what are, in effect, still the colonies. It remains a colonial relationship, with a veneer of good karma (Alyson, 2022).

Furthermore, Alyson (2022) affirmed that, to be clear, the UN and aid industry never call for rote-based schooling. Just the opposite. The "child-centered classroom" is one of their mantras. But in diverse countries around the world, in circumstances entirely different from what is familiar to top the UN officials, the one-size-fits-all of the United Nations' policies made this happen. In his writing:

> "Ill-prepared school systems had no other way to deal with the increasing enrollment numbers. Further, it was easily predictable, and was often reported as it took place. If good-hearted Joe guzzles half a bottle of gin, then speeds through town while singing a song about safe driving and kills three pedestrians, he is still a criminal."

He gave an example by a graduate of Teachers College at Columbia University, who has worked extensively in Asia and Africa, quoting Iyengar (2014):

> "My work takes me to very many different countries each time - Ghana, Nigeria, India, Malawi. These countries are very different in terms of the language spoken, food, culture …. everything. In the many years that I have been traveling, I have noticed a few common things. Visiting classrooms, we often see a teacher standing in front of the class and asking the children to repeat after him/her. Children dutifully repeating after the teacher, as some look outside the window curiously looking at the visitors… but the repeating doesn't stop. Very few textbooks and children sharing the few tattered textbooks they have. I usually carry some simple text with me just to see if they can read some simple materials. Many times, they are unable to. This unfortunately has become a very disturbing

trend in most developing countries. Children are in school, I can often hear loud chants of children repeating after their teachers, yet when I ask them to read big bold simple texts, they are unable to do so."

Again, quoting Economist (2017), it comes to knowledge that "often money causes the problem by attracting the wrong people. ... about failing schools in India reported that public schoolteachers often earn more than ten times the median local salary, yet "half of fifth-grade pupils (ten-year-old) cannot read a story designed for second-graders."

UNICEF (2022) feels concern about the contemporary global learning crisis: The global disruption to education caused by the COVD-19 pandemic is without parallel and the effects on learning are severe. The crisis brought education systems across the world to a halt, with school closures affecting more than 1.6 billion learners. While nearly every country in the world offered remote learning opportunities for students, the quality and reach of such initiatives varied greatly and were at best partial substitutes for in-person learning. It is more devastating in the Global South. Almost two years, schools remain closed for millions of children and youth, and millions more are at risk of never returning to school. Evidence of the detrimental impacts of school closures on children's learning offer a harrowing reality: learning losses are substantial, with the most marginalized children and youth often disproportionately affected (Alyson, 2022).

There are substantial differences in the kinds of challenges currently faced by education systems in the Global North and South. The pressures in the systems in the North is to compete to ensure more and more learners are succeeding in acquiring higher order learning skills as articulated in cross-national tests like Programme for International Student Assessment (PISA), Progress in International Reading Literacy Study (PIRLS) and Trends in International Mathematics and Science Study (TIMSS). At the level of instructional reform, these systems are finding ways to ensure that teachers are engaged in 'ambitious teaching.' In contrast, in the Global South, having only recently achieved universal school access are confronted with the problem that many children are in school but not learning to read, write, and become numerate. This is most clearly illustrated in the recent UNESCO global monitoring report that revealed that at least 250 million young people are failing to learn the basics, including a large proportion attending school (Rubin, 2022).

According to the World Bank (2019), one big reason the learning crisis persists is that many education systems across the developing world have little information on who is learning and who is not. As a result, it is hard for them to do anything about it. And with uncertainty about the kinds of skills the jobs of the future will require, schools and teachers must prepare students with more than basic reading and writing skills. Students need to be able to interpret information, form opinions, be creative, communicate well, collaborate, and be resilient. This is why its vision is for all children and youth to be learning and acquiring the skills they need to be productive, fulfilled, and involved citizens and workers. Its focus is on helping teachers at all levels become more effective in facilitating learning, improving technology for learning, strengthening management of schools and systems, while ensuring learners of all ages, from preschool to adulthood, are equipped for success.

Towards Sustainable Education for the 21st Century in the Global South

It has been demonstrated that education originated as transmission of cultural heritage from one generation to the next. At the moment, the goals are increasingly encompassing new ideas such as liberation of learners, critical thinking about presented information, skills needed for the modern society, empathy, and complex vocational skills. UNESCO believes that it is a human right for all throughout life and that access must be matched by quality (UNESCO, 2015, 2018). The organization has been entrusted to lead the Global Education 2030 Agenda through Sustainable Development Goal 4. The UNESCO (2018) outlines the objectives to include:

> Target 4.1. Free, equitable and quality primary and secondary education. By 2030, ensure that all children complete free, equitable and quality primary and secondary education leading to relevant and effective learning outcomes.
>
> Target 4.2. Quality early childhood development, care, and pre-primary education. By 2030, ensure that all children have access to quality early childhood development, care, and pre-primary education so that they are ready for primary education.
>
> Target 4.3. Quality TVET and tertiary education. By 2030, ensure equal access for all people to affordable and quality technical, vocational, and tertiary education, including university.
>
> Target 4.4. Technical and vocational skills. By 2030, substantially increase the number of youth and adults who have relevant skills, including technical and vocational skills, for employment, decent jobs, and entrepreneurship.
>
> Target 4.5. Equal access to all levels of education and training for the vulnerable. By 2030, eliminate gender disparities in education and ensure equal access to all levels of education and vocational training for the vulnerable, including persons with disabilities, indigenous peoples, and children in vulnerable situations.

> Target 4.6. Youth and adult literacy and numeracy. By 2030, ensure that all youth and a substantial proportion of adults, both men and women, achieve literacy and numeracy.
>
> Target 4.7. Knowledge and skills needed to promote sustainable development. By 2030, ensure that all learners acquire the knowledge and skills needed to promote sustainable development, including, among others, through education for sustainable development and sustainable lifestyles, human rights, gender equality, promotion of a culture of peace and non-violence, global citizenship, and appreciation of cultural diversity and of culture's contribution to sustainable development.
>
> 4.a. School environment. Build and upgrade education facilities that are child, disability, and gender sensitive and provide safe, non-violent, inclusive, and effective learning environments for all.
>
> 4.b. Scholarships. By 2020, substantially expand globally the number of scholarships available to developing countries, in particular least developed countries, small island developing States and African countries, for enrolment in higher education, including vocational training and information and communications technology, technical, engineering, and scientific programmes, in developed countries and other developing countries.
>
> 4.c. Teachers. By 2030, substantially increase the supply of qualified teachers, including through international cooperation for teacher training in developing countries, especially least developed countries and small island developing States.

If it should be justly the process of facilitating learning, or the acquisition of knowledge, skills, values, morals, beliefs, habits, and personal development (UNESCO, 2021), in the 21st Century, there must be radical departure from the rhetoric.

The emerging evidence from the Global South, as reflected in the experimental research from India, Kenya, and South Africa, is that combined and structured intervention programmes need to focus on early grade learning, particularly in areas of literacy in local languages and second language. These initiatives would be geared to change entrenched instructional practices and thereby impact positively on learning outcomes. In so many ways, at large-scale and otherwise, system-wide, the interventions would impact core elements of instruction - enhance teachers' instructional knowledge and skills, upgrade the educational materials available to learners and change the typical learning tasks children do in the classrooms (Rubin, 2022).

There are three sites in the Global South that point to positive change (Rubin, 2022): Large Indian not-for-profit organization, focused on high-quality, low-cost, and replicable interventions; the Primary Mathematics and

Reading Initiative (PRIMR) experience and national roll-out in over 22,000 schools; and the Primary Language and Mathematics Initiatives (GPLMS) and the Early Grade Reading Study (EGRS) in South Africa. Although each system context is making unique contribution to the field of educational change, they share in common a commitment to the use of rigorous research methods including experimental studies (Rubin, 2022).

Within the field of educational change, the focus has been on developed school systems (in North America and Europe), with growing interest in 'high performing' systems in East and Southeast Asia. The knowledge base in the field of educational change has largely been built on case studies of 'successful' districts, provinces, or countries. There have been emergence of new knowledge base from the Global South as represented in the experimental research on system-wide improvement of early grade learning. Unlike the methodological orientation in the field in the Global North, the research from the South is increasingly building on the accumulation of findings on robust models using large-scale randomized trials. This experimental research tradition is pointing to key stakeholders, international donors/funders, policy makers and system-managers, to what works in resource-constrained contexts with limited professional capital (Rubin, 2022).

The global development landscape has fundamentally changed allowing for new forms of partnership and cooperation to flourish (UNESCO, 2021). Traditionally, aid flowed from the developed North to the developing South. Now developing countries are building economic and other cooperation relationships between each other at an unprecedented rate. This represents the exchange of knowledge and resources between governments, organizations, and individuals in developing countries, or those from what is known as the Global South. This may be based on individual or shared development objectives. North-South-South Cooperation or Triangular Cooperation describes two or more developing countries collaborating with a third developed country which contributes knowledge, technology, and resources.

Emerging economies play an increasingly visible role in international education cooperation with a move from traditional transfers of money to technical and knowledge transfers. As the landscape has shifted, role of UNESCO has evolved. It has always promoted partnership, the sharing of information and best practices and networking as an essential feature of its work in education. As leader of the Education 2030 Agenda (and beyond), it is now more firmly placed to function as broker, instigator, and convenor of new and stronger partnership arrangements on a global, inter-regional, regional, or national scale.

UNESCO would have to continue forging equal partnerships between countries, in particular, by strengthening technical cooperation between developing countries and attracting funding from new and emerging donors.

It would continue to cooperate with Member States, civil society, and the academic world, and promotes public-private partnerships in education.

The pivotal functions of UNESCO in this role have included providing a common platform for engagement through international and regional fora, acting as a clearing house to share good practices in education, building networks and communities of practice and rewarding innovative and outstanding projects through prizes. Example of such projects promoting South cooperation have been: Capacity development (CapED Programme); Teacher education (Enhancing teacher education in Africa, Building the capacities of teachers and teacher trainers, and Enhancing teacher participation in social dialogue); Girls' and women's education (UNESCO Malala Fund for Girls' Right to Education, UNESCO-CJ Strategic Partnership for Girls' Education; UNESCO-HNA Partnership for Girls' and Women's Education, Empowering adolescent girls and young women through education), Technical and vocational education and training (The Better Education for Africa's Rise project, and Youth Employment in the Mediterranean (YEM)); ICT in education (ICT transforming education in Africa, Best practices in mobile learning, Leveraging ICT to achieve education 2030); Good practices (Effective Literacy and Numeracy Practices Database (LitBase); UNESCO HIV and Health Education Clearinghouse, IIEP Learning Portal); International and regional fora (Collective Consultation of NGOs on Education 2030, E-9 partnership, Education 2030 in Africa, Education 2030 in the Arab States); and Networks and knowledge exchange (UNESCO Global Network of Learning Cities, UNESCO Associated Schools Network, UNEVOC Network Portal, MGEIP clearing house on global citizenship education, and International Centre for Higher Education Innovation). They have all been realistic and encouraging.

Similarly, there have been prizes, namely: UNESCO International Literacy Prizes, UNESCO-Japan Prize on education for sustainable development, UNESCO Prize for girls' and women's education, UNESCO-Hamdan Bin Rashid Al-Maktoum Prize for outstanding practice and performance in enhancing the effectiveness of teachers, and UNESCO King Hamad Bin Isa Al Khalifa Prize for the use of information and communication technologies in education. The granting of prizes should be continued and intensified to align the community, and to remind everyone of ideals and virtues valued in the education sector. Each simply represents a democratization of sorts, in this instance, of the expression of those ideals and virtues (UNESCO, 2021b).

There have been successful efforts in rapidly changing nations and regions, with wide-ranging collection that grapples with accreditation, credentialing, quality standards, innovative assessment, learner motivation and attrition, and numerous other issues in eLearning, ICT, IT, and others. The provocative narratives curated in the collections have demonstrated how massive open online courses (MOOCs) and open educational resources (OER) can be effectively designed and implemented in vastly different ways in particular settings, as could be detailed by experts. It could be essential resource for policy makers, instructional designers, practitioners, administrators, and other MOOC and OER community stakeholders in the educational sector (Zhang et al., 2019).

With e-learning technologies evolving and expanding at high rates, institutions (and organizations) around the world are integrating MOOCs and other OERs. *MOOCs and Open Education in the Global South* must explore the initiatives and leverage on the flexible systems to educate, train, and empower populations previously denied access to such opportunities.

Following from the foregoing, the challenges of infrastructure - electricity and power, IT and ICT hardware and software, manpower development, security of lives and property, lack of basic literacy, digital divide and others require collaborative approach from all stakeholders for enduring and sustainable education. It should be perceived as high priority for all stakeholders. Social and economic instability, corruption, *I-Don-Care attitudes* of political officeholders, kidnapping, famer-Fulani mayhem, insurgencies, internal displacement of people, and others are directly and indirectly negatively impacting education at all levels. They are terribly threatening sustainable education of the 21st Century.

World leaders and their advisers did not display the wise tools needed to cope with the challenges probably since their education was provided in those silos (Targowski, 2011). This chapter would agree with Targowski (2011) that suggests that education in the 21st Century must consider the following factors:

- Profound innovations in technology which change the modus operandi of society, for example, transforming it from real to virtual. The latter forms a new political elite of 100 million, mostly young people who are in touch and informed instantly, living in all corners of the world.
- The fast spread of knowledge from advanced to developing nations. In effect, China became the World Factory and India became the World Laboratory, and the developed nations are losing their middle classes and their purchasing power.
- The growth of population and its concentration in cities as well as international migration destabilizes local labor forces.
- The financial integration of the world which leads to huge wealth interception by a very narrow group of people under the flag of the Managerial Revolution.
- The declining morale and ethics among business elite of the world.

- Globalization which threatens localization and triggers issues of supranational versus subnational problems and challenges.
- The rising demands for political and human rights.
- The clash of the Islamic Civilization with Western and Eastern Civilizations.
- The limits of capitalism, socialism, and communism as effective and socially acceptable world-systems and the lack of new ideas.
- The fast demise of world reserves of strategic resources.
- Diminishing environment due to overpopulation and industrialization, and lack of effective national and international counter measures; and
- The idealistic spread of democracy which is powerless in solving all these problems.

These factors are subjects of education but not in a consolidated manner or even in a formal manner as a legitimate part of curricula. Furthermore, the faculties are reluctant in changing ways and contents of teaching, due to routine and easiness in carrying their duties; and the administration in higher education, particularly in the Western Civilization, is detached from academic processes, seeing its role as "playing safe," and looking for the next better paid job elsewhere (Targowski, 2011).

Participants at an American Council on International Intercultural Education (ACIIE) conference adopted a framework for global education. The participants defined a competency as "an ability, a skill, some knowledge, or an attitude that can be demonstrated, observed or measured" and includes cognitive as well as affective attributes (ACIIE, 1996:3). The conferees created an acceptable list of developmental stages that "represent a continuum germane throughout all levels of education," namely:

- Recognition of global systems and their connectedness, including personal awareness and openness to other cultures, values, and attitudes at home and abroad.
- Intercultural skills and direct experiences.
- General knowledge of history and world events - politics, economics, geography, and others; and
- Detailed area studies specialization - expertise in another language, culture, or country.

The following competencies were selected as the most important among the fifty generated by the participants: The globally competent learner is empowered by the experience of global education to help make a difference in society; committed to lifelong, global learning; aware of diversity, commonalities, and interdependence; recognizes the geopolitical and economic interdependence of the world; appreciates the impact of other cultures;

accepts the importance of all peoples; capable of working in diverse teams; understand the non-universality of culture, religion, and values; and accepts responsibility for global citizenship (Targowski, 2011).

A review of literature suggests that these competencies are being challenged by the emergence of new pedagogical techniques, necessary changes in programme implementation, evolving theoretical perspectives, and the educational imperativeness to develop critical thinking of students (versus unacceptable memorization). Therefore, one must emphasize that "Global education is an imperative not an option" if the education and graduates are to survive globalization (American Council on International and Intercultural Education [American Association of Community Colleges] (ACIIE), 1996:2). According to Zeszotarski (2001) "ACIIE presents a rather narrow conception of globalization that does not acknowledge the cultural, political, or social dimensions of internationalization." Much of the public and academic debate on globalization focuses on issues surrounding the rise of the global economy: the worldwide influence of capitalism, the rise of information as the significant economic product, the need for knowledgeable workers skilled in information technology, and the new role of labor. Despite the focus on economic forces, globalization has a social and cultural face as well. The spread of electronic communications technology in particular has had a profound effect on developing what McLuhan (1962 & 1996) called the "global village." The current level of education in the 21st Century is mostly offered at the level of small-picture and in the realm of unlimited strategic resources, which could be acceptable in the 19th and 20th centuries but not in the 21st Century. At the level of the big-picture of civilization, the state of the world is more pessimistic than optimistic as it was in the past two centuries, when "the sky was the limit." But the sky is not the limit anymore (Targowski, 2011).

Notwithstanding the great progress in primary schooling in some developing countries, the preparation of youth for work and life is very low, just as demand for knowledge and skills is rising. Past education policies focused on increasing the number of people who go through the education system rather than teaching and learning the "right stuff" in schools. As a result, there are many pseudo-educated people who cannot read and write properly even in developed nations. To prepare young people for future work and life, they must learn the "right stuff" (Targowski, 2011). Education opportunities are not enough when young people and the society cannot benefit from them. In order to be successful in life, young people should have a solid foundation preparing them for post-primary education. This foundation is at the level of efficient knowledge, skills, and wisdom, such as (Targowski, 2011) suggests:

Basic knowledge and skills:

- Special knowledge (including numeracy and literacy)
- Life on Earth (sustainable eco-system and common goods)
- Life in the Society (end of life, virtues, and values)

Post-basic knowledge:

- Thinking skills (critical, creative thinking, and problem-solving)
- Higher order behavioural skills (decision-making, perseverance, self-discipline, teamwork, the ability to negotiate conflict and manage risk)
- Life-long learning attitude Life in Civilization (cross-culture and cross-civilization interaction)

Specific knowledge, skills, and wisdom :

- Vocational knowledge and skills (to perform jobs that rely on clearly defined tasks)
- Prudent judgment of real-life situations (adapted in its complexity to each level of education).

The presented set of knowledge, wisdom and skills is only partially applicable at the level of primary and secondary education, even in the developed countries. One of the reasons is the lack of properly educated teachers and education leaders. Hence, the World Civilization in the 21st Century is misguided almost at all levels of the society. Its revival without properly educated "global citizens" is not possible (Targowski, 2011).

CONCLUSION

The educational development in the Global South follows closely the development in the Global North. But the Global South is lagging far behind whereas the demand for sustainable education is ubiquitously required for sustainable living anywhere around the globe. The challenges militating against attainment of sustainable education require holistic approach as reiterated in this chapter. In addition, the followings are imperative and should be reiterated here: lack of infrastructure; political wills on the parts of public and political office holders; human and material development; inclusive education; security of lives and properties; and funding. These are in no small ways significant for sustainable education; and must be squarely addressed by every stakeholder.

This chapter opines that inclusive education whereby all students of all capabilities have the opportunity to grow and learn is aligned with the curriculum of schools and colleges (private and public). This means providing each student (and parent) with access to accurate information on their learning through ongoing formative and summative assessment of the progress of each student. But, if we are placing concepts created in Western Europe and North America, or what Connell calls the 'global metropole' at the heart of global agendas and goals (particularly referred in this case to the use of the term 'inclusive education' in the 4th SGD), it will be prudent to remember that such concepts are likely to acquire different meanings in the contexts of the Global South. For this reason, it is important to continue to learn more from so far neglected Global South contexts about the harsh realities, anxieties and obstacles people face in the complex spaces. More culturally sensitive research is needed on inclusive education in the Global South, because in the current global climate of increased efforts to ensure participation of all children in education, there is great need to approach local and cultural contexts that still remain largely unexplored. Like in Colombia, many communities in the Global South space are still in the process of negotiating their own understandings of inclusive education and the local relevance of the Global North-created concept.

The contemporary oversimplified focus on one category of need is harmful for most of the Global South contexts, because it renders invisible the plethora of vulnerable children that live in these contexts and leaves the complex intersections between factors leading to educational disadvantaged unchallenged and unresolved. As a way forward, the Global South needs time to reflect on the meaning of inclusive education within its current historical and cultural context, and to plan teacher preparation and practice in a way that will be both meaningful and sustainable.

At closer range, UNESCO should continue to provide policy-makers with an authoritative data, information, and directions with which to advocate that education be a cornerstone of the 21st Century global development architecture. The EFA Global Monitoring Reports have been found to be of relevance to all stakeholders. It should continue to be editorially independent and evidence-based publication that would serve as an indispensable tool for governments, researchers, education and development specialists, media, and students; and it should be made available to policy makers in the educational sectors of all nations. Since 2002, it has assessed education progress in some 200 countries and territories on an almost annual basis.

UNESCO believes that education is a human right for all throughout life and that access must be matched by quality. The Organization is the only United Nations agency with a mandate to cover all aspects of education. As noted earlier, it has been entrusted to lead the Global Education 2030 Agenda through Sustainable Development Goal 4. The Global South will benefit immensely by dependent on it as it has so much to offer for progress in educational development.

The vision of the World Bank is for all children and youth to be learning and acquiring the skills they need to be productive, fulfilled, and involved citizens and workers. Its focus in helping teachers at all levels become more effective in facilitating learning, improving technology for learning, strengthening management of schools and systems, while ensuring learners of all ages, from preschool to adulthood, are equipped for success. These should continue and with emphasis in the Global South.

REFERENCES

Abadzi, H. (2017). Turning a molehill into a mountain? How reading curricula are failing the poor worldwide. *Prospects* 46, 319-334. https://doi.org/10.1007/s11125-017-9394-9

Al-Hassani, S. T. S. (2011). *1001 inventions: Muslim heritage in our world.* Foundation for Science, Technology and Civilization Ltd.

Alyson, S. (2022 January 17). Schools in the global South are getting worse. We need to ask why. UNESCO Institute for Statistics. https://karmacolonialism.org/schools-in-the-global-south-are-getting-worse-we-need-to-ask-why/.

American Council on International Intercultural Education. (1996, November). Educating for the global community: A framework for community colleges. Report of a conference sponsored by the American Council on International Intercultural Education and the Stanley Foundation (Warrenton, VA.) (ED 407 958).

Arong, E. & Ogbadu, M. A. (2010). Major Causes of Declining Quality of Education in Nigeria from Administrative Perspective, *Canadian Social Science, 6*(3), 183-198.

Baines, J. (1983). Literacy and ancient Egyptian society, *Man (New Series), 18*(3), 572-599.

Baker, E., Peterson, P. & McGaw, B. (Eds.) (2010). *International Encyclopedia of Education* (3rd ed. 8 vol). Elsevier Science Ltd.

Benavot, A., & Julia, R. (2006). Lessons from the past: A comparative socio-historical analysis of primary and secondary education. In J. Colton et al. (Eds.) *Educating all children: A global agenda*: 123-229. Online

Bowen, J. (1981). *A History of Western Education (3 vol.) The modern West: Europe and the New World.* Methuen, 631 pp.

Compayre, G. & Payne, W. H. (2003). *History of Pedagogy.* Kessinger Publishing, p. 9.

Creffier, M. (2007 May 11). More in School, But Not Learning Enough, Inter Press Service News Agency. https://www.ipsnews.net/2007/05/development-more-in-school-but-not-learning-enough/.

Dados, N. & Connell, R. (2012). The Global South. *Key Concepts in Social Research, 11*(1), 12-13.

Education.govt.nz. (2020 October 24). Practical information about education for parents and careers. Education.govt.nz.https://parents.education.govt.nz/secondary -school/your-child-at -school /leaving-school-before-the-age-of-16/.

Fischer, S. R. (2004). *A History of Writing*, Reaktion Books, p. 50.

Gate, B. (2018 January 15). The Optimists. *Time* magazine. 191(1). https://time.com /magazine/ us /5087338/january-15th-2018-vol-191-no-1-u-s/.

Global Partnership for Education (GPE). (2021). 2021: GPE Results Report 2021: Final results report on GPE's 2016-2020 strategy, *Global Partnership for Education.* https://www. globalpartnership.org/sites/default/files/docs/results-report-202 1/en/2021-10-GPE-Results-Report-2021-v2.pdf.

Gross-Loh, C. (2014 March). Finnish Education Chief: 'We Created a School System Based on Equality' An interview with the country's minister of education, Krista Kiuru. The Atlantic. https://www.theatlantic.com/education/archive /2014/03/finnish-education-chief-we-created-a-school-system-based-on-equality /284427/.

Harris, R. (2003). *Gender and aging in Mesopotamia.* University of Oklahoma Press, 288 pp.

Hollington, A., Tappe, O., Salverda, T. & Schwarz, T. (2021 November 21). *Introduction: Concept of the Global South.* Global South Study Center. Cologne. https://web.archive. org/web/20160904205139/http://gssc.uni-koeln.de/node /451.

Hopkins, K. (1991). Conquest by book. In J. H. Humphrey (Ed.) *Literacy in the Roman World: Journal of Roman Archeology, Supplementary Series*, University of Michigan, Ann Arbor, Mich, *No 3*, pp. 133-158.

Iyengar, R. (2014 October 27). Back to Basics: Tackling Low-Levels of Learning in Chichewa, Millennium Villages. https://web.archive.org/web/20150214215652 / http://millennium villages.org/field-notes/back-to-basics-tackling-low-levels-of-le arning-in-chichewa/.

Kagan, D., Ozment, S. & Turner, F. M. (2007). *Western Heritage: Since 1300* (AP Edition). Pearson Education, Inc.

Kayagil, A., Hill, T & García, D. R. (2018 February 22). Barriers to Education Reform in the Global South: Teachers Unions. https://docs.google.com/document /d/1AJVs Agd3 hdPtCnVNU8 vuqOb7TINJMWhHuWhHEP7dqyo/edit.

Lin, J. (1993). *Education in Post-Mao China.* Praeger, 408 pp. https://www.bing.com /search? q=Education+in+Post-Mao+China&cvid=7ca5718ad5de437789bde372d cac4589&aqs= edge..69i57j0.2338j0j1&pglt=43&FORM=ANNTA1&PC=LCTS 1993).

Lu, J. (2017 September, 29). More and more children are in school around the world, But they are not learning much. UN Dispatch (United Nations News and Commentary – Global News Commentary. https://www.undispatch.com/children-school -around-world-not-learning-much/#:~:text=More%20and%20More%20Children %20are%20in%20School% 20Around,battle.%20Making%20sure%20they% E2%80%99re%20learning%20is%20also%20critical.

Lumenwaymaker. (n.d.). *Introduction to Sociology: Education Around the World.* Lumenwaymaker. https://courses.lumenlearning.com/wm-introductiontosociology /chapter/education-around -the-world/.

Mahler, A. G. (2018). *From the Tricontinental to the Global South: Race, Radicalism, and Transnational Solidarity*. Duke University Press.

McCreadie, M. (2012). *The Evolution of Education in Australia.* IFHAA Australian Schools.

McCulloch, G. & Crook, D. (Eds.) (2013). *The Routledge International Encyclopedia of Education.* Routledge.

McGovern Dole International Food for Education and Child Nutrition Program (2015). FY 2016 Food Assistance Proposal Guidance and Request for Applications, United States Department of Agriculture (USDA), Foreign Agricultural Service.

McIntosh, M. A. (2021 June 23). A History of Education since Ancient Civilizations. https:// brewminate.com/a-history-of-education-since-ancient-civilizations/.

McLuhan, E. (1962). *The Global Village.* Living Internet. https://www.livinginternet. com/i/ii mcluhan.htm#:~:text=Marshall%20McLuhan%20predicted%20the%20g lobal%20village%2C%20one%20world,global%20village%20and%20to%20co nsider%20its%20social%20effects.

McLuhan, E. (1996). *The Source of the Term 'Global Village'.* McLuhan Studies. 2, pp. 157-158.

Milian, C. (2013). Southern Latinities. *Latining America: Black-Brown Passages and the Coloring of Latino/a Studies*. Athens: University of Georgia Press.

National Center for Public Policy and Higher Education. (2006). Measuring UP: The National Report Card on Higher Education. http://www.eric.ed.gov/PDFS/ ED493360.pdf.

National Public Radio. (2010). Study Confirms U.S. Falling Behind in Education. All Things Considered, December 10. https://www.npr.org/2010/12/07/131884477/ Study-Confirms-U-S-Falling-Behind-In-Education.

Oloruntimehin, B. O. (1974). Education for Colonial Dominance in French West Africa from 1900 to the Second World War. *Journal of the Historical Society of Nigeria 7*(2), 347-356.

OpenStax. (2022). Education Around the World. OpenStax, Rice University. https:// openstax. org/books/introduction-sociology-3e/pages/16-1-education-around-t he-world.

Otieno, M. O. (2019 March 21). *What is the Global South?* World Facts. https://www. worldatlas.com/articles/what-is-the-global-south.html.

Paglayan, A. S. (2021). The Non-Democratic Roots of Mass Education: Evidence from 200 Years. *American Political Science Review 115*(1), 179-198.

Parsons, M. (2022). *Education in Ancient Egypt.* Tour Egypt. http://www.touregypt .net/ feature stories/educate.htm.

Pellissier, H. (2013 October 23). High test scores, higher expectations, and presidential hype: Should South Korean schools point the way to American reforms? Great Scholar.org. https://www.greatschools.org/gk/articles/south-korean-schools/#:~ :text=South%20 Koreans %20attend%20school%20220%20days%20per%20 year%2C,ceased%20 going %20to%20school%20half%20days%20on%20Saturday.%29.

Prashad, V. (2012). *The Poorer Nations: A Possible History of the Global South.* Verso.

Programme for International Student Assessment (PISA) (2015). Results in Focus. Programme for International Student Assessment. https://www.oecd.org/pisa/pisa-2015-results-in-focus.pdf.

Rowley, K. J., Silveira, F., Dufur, M. J. & Jarvis. J. A. (2019). U.S. Underperformance in an International Context. *Teachers College Record.* 121(10), 1-46.

Rubin, C. M. (2022). The Global Search for Education: Impact of Globalization on the North and South Divide. *Education World: Connecting Education to What Works.* https://www.Educ ationworld.com/blog/global-search-education-impact-globalization-north-and-south-divide.

Targowski, A. (2011). The Civilization Approach to Education in the 21st Century, *Comparative Civilizations Review. 65*((9), 73-104. Available at: https://scholarsarchive.byu. edu/ccr /vol65/iss65/9.

The Economist. (2012 June 16). Education in Thailand: Let them eat tablet. The Economist. https://www.economist.com/asia/2012/06/16/let-them-eat-tablets.

The Economist. (2017 June 10). Indian education: Why the world's biggest school system is failing its pupils. The Economist. https://www.economist.com/asia/2017/06/08/why-the-worlds-biggest-school-system-is-failing-its-pupils.

The World Bank. (2021). Literacy rate, adult total (% of people ages 15 and above). Retrieved from. The World Bank. https://data.worldbank.org/indicator/se.adt.litr.zs.

Thomason, A. K. (2005). *Luxury and Legitimation: Royal Collecting in Ancient Mesopotamia.* Ashgate Publishing Ltd., p. 25.

Time Magazine. (2014 July 8). More Children Are Going to School in India, but They're Learning Less. Time Magazine. https://time.com/2956888/india-education-decline-world-bank-report/.

Ukaigwe, P.C. & Chima, D. N. (2019). Comparative analysis of budgetary allocation to education and life expectancy in Nigeria, 1980-2015. *International Journal of Education, Learning and Development,* 7(4), pp.84-96.

UNESCO Institute for Statistics. (2018). Quick Guide to Education Indicators for SDG 4. UNESCO Institute for Statistics. UIS/2018/ED/TD/4. http://uis.unesco.org/sites/ default /files/documents/quick-guide-education-indicators-sdg4-2018-en.pdf#:~:text=SDG%204 %20aims%20to%20%E2%80%9Censure%20inclusive%20and%20equitable,a%20sustainable%20education%20agenda.%20What%20is%20a%20target%3F.

United Nations Children's Fund (UNICEF). (2020). Global learning crisis. UNICEF. https://www. unicef.org/reports/state-global-education-crisis

United Nations Children's Fund (UNICEF). (2021 December). *The State of the Global Education Crisis: A Path to Recovery* charts a path out of the global education crisis and towards building more effective, equitable and resilient education systems. UNICEF. https://www. unicef.org/reports/state-global-education-crisis.

United Nations Children's Fund (UNICEF). (2019) Every Child Learns: UNICEF Education Strategy 2019–2030. UNICEF. https://www.unicef.org/reports/UNICEF-education-strategy-2019-2030.

United Nations Educational Scientific and Cultural Organization (UNESCO). (2022 January 17). Schools in the global South are getting worse. We need to ask why. UNESCO Institute for Statistics.https://karmacolonialism.org/schools-in-the-global-south-are-getting-worse-we- need-to-ask-why//

United Nations Educational Scientific and Cultural Organization (UNESCO). (2017a September 21). More than one-half of children and adolescents are not learning worldwide. UNESCO Institute of Statistics. http://uis.unesco.org/en/news/6-out-10-children-and-adolescents-are-not-learning-minimum-reading-and-math#:~:text=Globally%2C%20six%20 out%20 of %20ten%20children%20and%20adolescents,230%20million%20adolescents%20of%20 lower%20secondary%20school%20 age.

United Nations Educational Scientific and Cultural Organization (UNESCO). (2015). Education for All 2000-2015: Achievements and Challenges. UNESCO. https://books.google. com/ books/about/Education_for_All_2000_2015_Achievements.html?id=T-PvCAAAQBAJ #:~ :text=Education%20for%20All%202000---2015%3A%20 Achievements%20and% 20 Challenges%20provides,for%20monitoring%20and%20evaluating%20education%20 targets%20after%202015.

United Nations Educational Scientific and Cultural Organization (UNESCO). (2021a). South Cooperation in Education. UNESCO. https://en.unesco.org/themes/education/ south-south-cooperation

United Nations Educational, Scientific and Cultural Organization (UNESCO). (2018 August 28). The Learning Crisis is Causing a Skills Crisis. Here's Why. UNESCO Institute for Statistics. http://uis.unesco.org/en/blog/learning-crisis-causing-skills-crisis-heres-why#:~: text=According%20to%20data%20from%20the%20UNESCO%20Institute%20for,school%2C%20or%20were%20in%20school%2C%20but%20dropped%20out

United Nations Educational, Scientific and Cultural Organization (UNESCO). (2017b). More than one-half of children and adolescents are not learning world-wide. UNESCO Institute for Statistics. http://uis.unesco.org/sites/default/files/documents/fs46-more-than-half-children -not-learning-en-2017.pdf.

United Nations Educational, Scientific and Cultural Organization (UNESCO). (2021c). SDG Resources for Educators - Quality Education. UNESCO. https://web.archive.org/web/ 20220 121083949/https://en.unesco.org/themes/education/sdgs/material/04.

United Nations Educational, Scientific and Cultural Organization (UNESCO). (2021b). UNESCO South-South Cooperation in Education. UNESCO. https://en.unesco.org/themes /education/south-south-cooperation.

United Nations Finance Center for South-South Cooperation (FCFSSC). (2022). *Global South Nation.* United Nations FCFSSC.

West-Pavlov, N. & Connell, R. (2018). *The Global South and Literature*. Cambridge University Press.

World Bank. (2007). *World Bank Global Monitoring Report 2007*. World Bank.

World Bank. (2019 January 22). The Education Crisis: Being in School Is Not the Same as Learning. World Bank. https://www.worldbank.org/en/news/immersive-story/ 2019/01/22/ pass-or-fail-how-can-the-world-do-its-homework#:~:text=A%20global%20learning%20 crisis%20One%20big%20reason%20the,hard%20for%20them%20to%20do%20anything%20about%20it.

Yuan, Z. (1994). Local Government Schools in Sung China: A Reassessment, *History of Education Quarterly, 34*(2), 193-213.

Zhang, K., Bonk, C. J., Reeves, T. C. & Reynolds, T. H. (2019 November 13). MOOCs and Open Education in the Global South: Challenges, Successes, and Opportunities (1st Edition). Routledge. 392pp.

Zeszotarski, P. (2001, Summer). Eric Review: Issues in global education initiatives in the community college. *Community College Review, 29*(1), 6.

Chapter 7

Climate Change, Cross-Border Trade and Business Survival

By Sylvester Udeorah

INTRODUCTION

The issue of climate change has remained at the forefront of policy debates in recent times. This is because of the growing implications of changing climatic conditions and weather events on agricultural outputs, livelihood outcomes and gross domestic output (GDP). The International Panel on Climate Change (IPCC, 2014) describes climate change as variations that occur in the statistical distribution of weather over extended periods, typically ranging from decades to millions of years. These variations may occur in the average weather or simply in the distribution of weather events around an average, and may be limited to a particular region, or occurring across the globe. The Kuznet environment curve (KEC) hypothesis proposed that at the initial stage of economic development, higher income level initially have adverse effect on the environment and this adverse effect turn favourable as the country advances in the process of economic growth. The major contributors to climate are human activities through the use of nonrenewable energy supplies. According to the German Council on Climate Change (2013), the variability of climatic conditions poses a threat to human welfare and the natural ecosystem in both developed and developing economies. In comparison, Ajibade, and Eche (2017) posit that the effect of climate change on environmental sustainability is less severe in developed economies due to improved adaptation techniques, technological advancement, endowed advantage, and income status.

In addition, the International Panel on Climate Change (IPCC, 2007), predicted that countries in the sub-Saharan Africa are likely to be more vulnerable to climate change with projected agriculture loss of 2 to 7 of the countries' GDP. The report further explained that human activities altered the composition of the atmosphere, caused significant changes in the global climatic conditions. More recently, the 2021 UN Climate Change Conference (COP26) the panelists reached a consensus that Africa is particularly vulnerable to the risks posed by climate change, as the continent is experiencing increasingly frequent and severe climate shocks. Consequently, the African Group of Negotiators has continued to advocate for significant improvement in financial commitments of developed Countries to help African countries achieve their Nationally Determined Contributions (NDCs). Ding and Beh (2022) argue that the variability in climatic conditions and changing weather events have increased atmospheric heat while endangering the people's life through food insecurity and population displacement.

Climate change effects and implications on aquaculture production sustainability are expected to be both negative and positive although the adverse effects outweigh the positive ones. It is noteworthy that African nations have, in recent times experienced some unfavorable impacts of environmental change because of the elements of drought, rainfall, relative humidity and temperature amongst others. Emete and Amusa (2010) outlined the post-primary effects of climate change in Africa to include, including spread of pests, rising competition for available resources, and extinction of biodiversity. Kandu (2017) argues that climate change the predicted effects of climate change are increasingly threatening food security. Given the growing implications of climate change on trade competitiveness, agriculture development, and economic growth, adaptation strategies are constantly being integrated into the policy actions of achieving the Sustainable Development Goals (SGDs). It is important to note that investments in agriculture and environmental sustainability under climate change have been identified in economic literature as key to adaptations to climate change conditions. However, Huang and Wang (2014) opine that the commitment made at both global and national levels seems inadequate to engender efficient adaptations to climate change scenarios. More importantly, the policy environment required the development of specific adaptation strategies in view of the growing climate change, which seems to be sub-optimal in developing economies. Thus, this chapter deepened the understanding of the implications of climate change on trade and business survival.

CLIMATE CHANGE AND INTERNATIONAL TRADE

The environment and cross-border trade relationship is one of the keenly debated and research topics in contemporary environmental discuss. The level of consciousness and campaign is driven largely by the need to sustain the environment or planet for future generations and increasingly motivated by the negative consequences of climate change. The literature is split on the causal relationship between climate change and cross border or international trade. A strand of literature which favours globalization argue that trade-induced economic growth has the advantage of fostering cleaner environment for individuals with higher income, resulting from increased cross border trade, demand for a friendlier and cleaner environment. The result of this increased demand from both the citizens and employees of the manufacturing companies for environmentally friendly practices is a switch in production process that reduces the carbon footprint in the atmosphere. The other divide argues that increased production due to global trade depletes natural resources and increases pollution and environmental degradation which consequently deteriorates the quality of the environment.

In the last few decades, the world has experienced substantial changes in weather conditions and climate. These changes in climate have been traced to continued greenhouse gases (GHG) emissions which results from the activities of corporations and firms spread across the globe. The emissions of these greenhouse gases is more pronounced in the developed world as against the developing countries due to the nature of activities conducted. The production process in the developed countries employs heavy machineries which are emitters of greenhouse gases (GHG) such as carbon dioxide (CO2). The production activities in the developing or emerging economies is largely driven by more labor and less capital, hence they are categorized as less emitters of greenhouse gases (GHG) compared to their counterpart in the developed countries. Though the developed and developing countries differ in the volume of greenhouse gases, it is argued that developing countries are more likely to be susceptible to adverse climate and environmental change effects (IPCC, 2014).

According to Raza, Sui, Jermsittiparsert, Żukiewicz-Sobczak and Sobczak (2021), the effects of climate change are not uniform, and this has raised concern regarding climate change. Greenhouse gases which include carbon dioxide emissions, nitrous oxide, and methane emissions, are considered the drivers of climate change and the developed and developing countries differ in terms of the volume of greenhouse gases emitted. In the past one and half decades, there has been increase in the volume of carbon dioxide, methane, and nitrous oxide emissions across the globe. The trajectory of these greenhouse gas (GHG) emissions which manifest in the form of higher global surface temperatures and change in weather patterns are presented in Figures 7.1 through 7.4:

As observed from the diagrams above, carbon dioxide emissions witnessed an upward movement throughout the period. This suggests the increasing levels of climate change. Figure 7.1 revealed that carbon dioxide emissions across the world increased over the years. From 1981 to 2018, carbon dioxide

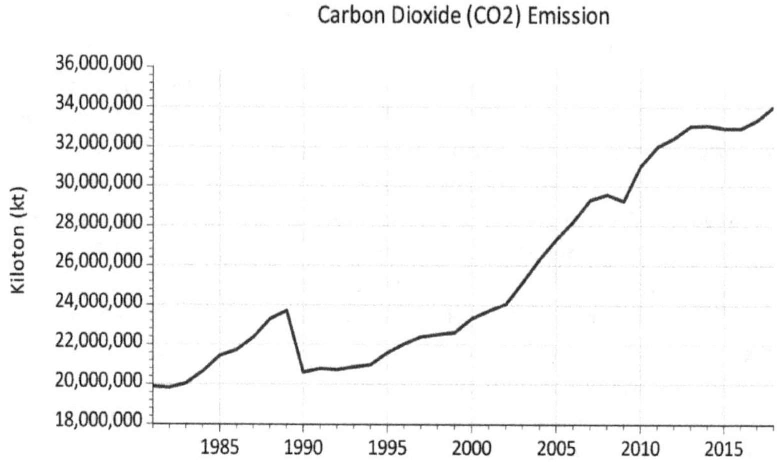

Figure 7.1 Trajectory of Global Carbon Dioxide (CO2) Emissions (1981–2018). https://databank.worldbank.org/source/world-development-indicators

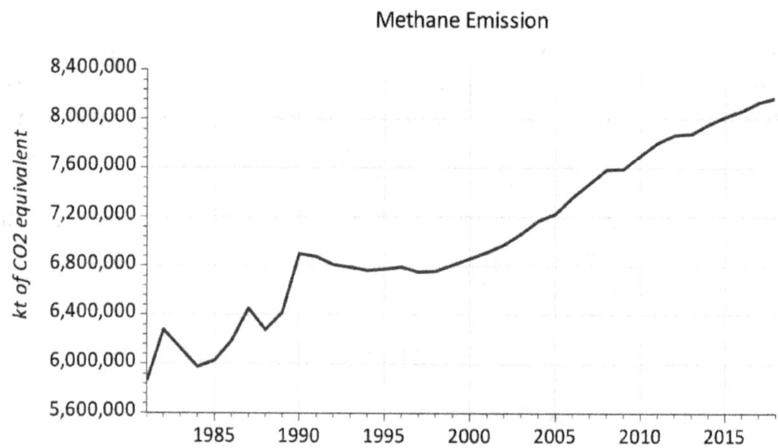

Figure 7.2 Trajectory of Global Methane Emissions (1981–2018). https://databank.worldbank.org/source/world-development-indicators

emissions averaged approximately 25.5 million kiloton and increased from 19.8 million kiloton (kt) to 34.0 million kt. This upward movement shows that there has been large concentration of carbon dioxide in the atmosphere and this gas, which is carbon dioxide, is one of the contributors to global warming and climate change. Similarly, trend was observed in Figure 7.2, which shows an upward trend in methane emissions from 1981 to 2018. Methane emissions during the period averaged 7.04 million kt. The minimum and maximum methane emissions were 5.86 and 8.17, respectively. With a

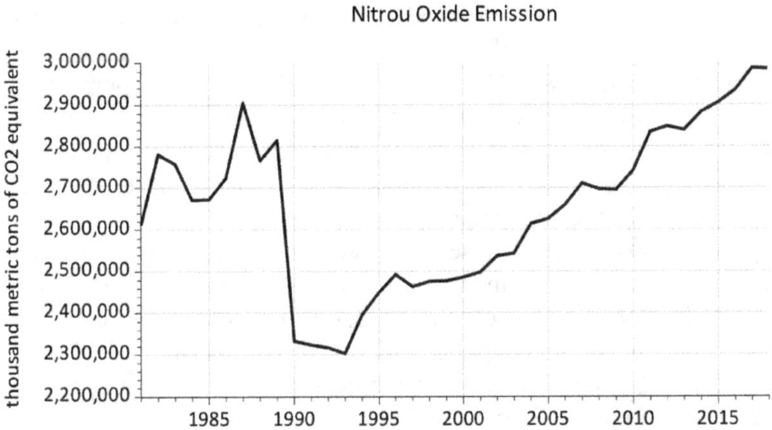

Figure 7.3 Trajectory of Global Nitrous Oxide Emissions (1981–2018). https://databank.worldbank.org/source/world-development-indicators

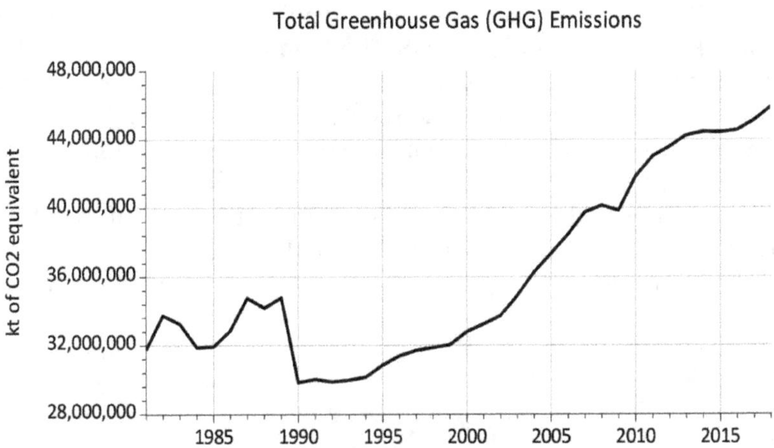

Figure 7.4 Trajectory of Global Greenhouse Gases (GHG) Emissions (1981–2018). https://databank.worldbank.org/source/world-development-indicators

skewness of 0.117270, this affirms that global methane emissions have been increasing over the period of study. There was significant increase in nitrous oxide emissions between 1981 to 1988 as reported in Figure 7.3. Global nitrous oxide emissions rose from 2.61 million kt in 1988 to 2.30 million kt in 1993. Afterward, global nitrous oxide emissions increased from 2.39 million kt in 1994 to 2.98 million kt in 2018.

Total greenhouse gases (GHG) emissions have not slowed down but continued to increase globally despite measures adopted by countries to reduce carbon footprint and go green, thereby reducing the emissions of GHG into the atmosphere, thereby ensuring the effects of climate change are lesser. As revealed by Figure 7.4, total GHG increased from 3.17 million kt in 1981 to 3.48 million kt in 1989. The emissions of GHG drop temporarily to 2.98 million kt in 1990, before rising to 4.59 million kt in 2018. Overall, Figures 7.1 through 7.4 seems to contradict the environmental Kuznets curve (EKC) or the inverted U hypothesis as it could be seen that at an advance level of economic development depicted by time of 2010 to 2018 which is the 21st century, emissions of greenhouse gases such as carbon dioxide, methane and nitrous oxide is still on the increase. Aside from the various climatic conditions, cross-border trade has continued to vary in recent times as reported in the trajectories of imports, exports, and global trade in Figures 7.5-7.9.

Figures 7.5 through 7.7 show the trajectories of global exports, global imports, and global trade from 1981 to 2020. Figures 7.5 shows that global exports and imports have been increasing over the years. Global export rose from US $2.02 million in 1981 to US $1.62 million in 2008. With shocks from the global financial crisis of 2008, global exports declined to US $1.26 million in 2009. Global exports fluctuated between 2014 and 2020 dropping from US $1.90 million in 2014 to US $1.60 million in 2016. This decline in export could be due to the crude oil crisis of 2014 – 2016 as global crude oil price fell to US $25 per barrel. The shocks from the coronavirus pandemic led to a decline in global export from US $19 million in 2019 to US $17 million in 2020.

Movement in global imports followed a similar pattern as those of global exports. Global imports rose from US $2.08 million in 1981 to US $16.5 million in 2008. The aftermath of the global financial crisis resulted in a decline

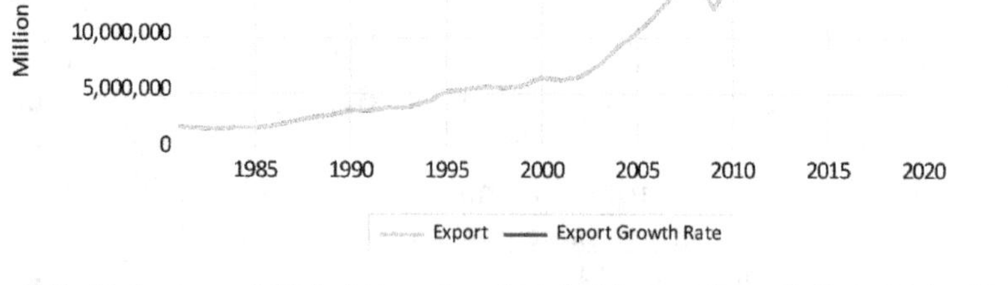

Figure 7.5 Trajectory of Global Export and Global Export Growth Rate. United Nations Conference on Trade and Development. (2021). Trade and Development Report 2021: From Recovery to Resilience – The Development Dimension. United Nations. Available at: https://unctad.org/system/files/official-document/tdr2021_en.pdf

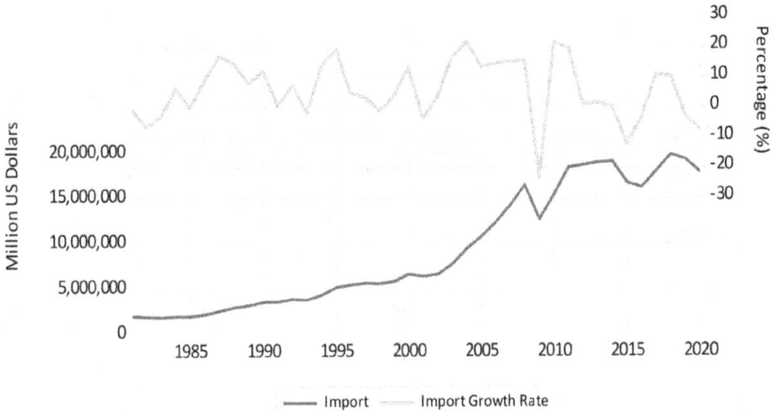

Figure 7.6 Trajectory of Global Imports and Global Import Growth Rate. United Nations Conference on Trade and Development. (2021). Trade and Development Report 2021: From Recovery to Resilience – The Development Dimension. United Nations. Available at: https://unctad.org/system/files/official-document/tdr2021_en.pdf

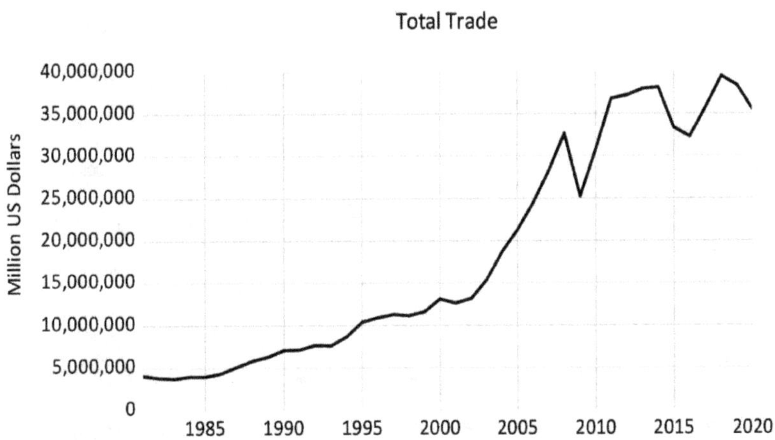

Figure 7.7 Trajectory of Global Trade. United Nations Conference on Trade and Development. (2021). Trade and Development Report 2021: From Recovery to Resilience – The Development Dimension. United Nations. Available at: https://unctad.org/system/files/official-document/tdr2021_en.pdf

in global import to US $12.7 million from the previous volume of import in 2008. Due to measures implemented to forestall the spread of the coronavirus, global import declined from US $19.3 million in 2019 to US $17.8 million in 2020 as reported in Figure 7.6. Cross border trade or global trade has increased over the period, only slowing downward trend during 2008 – 2009, 2014 – 2016, and 2019 – 2020, with each period decline due to shocks

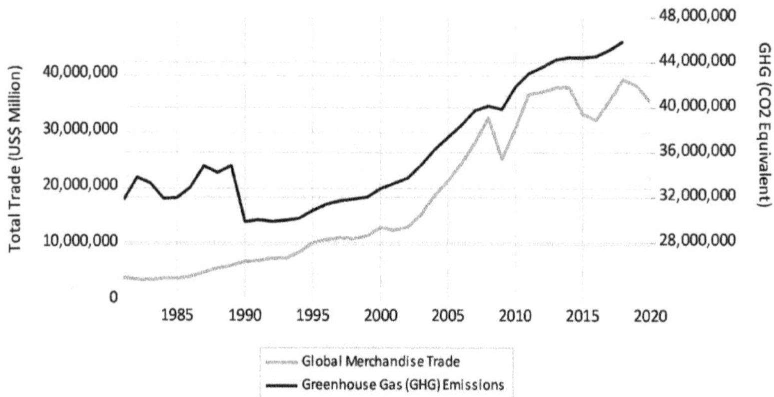

Figure 7.8 Trajectory of Global Merchandise Trade and Greenhouse Gas (GHG) Emissions. United Nations Conference on Trade and Development. (2021). Trade and Development Report 2021: From Recovery to Resilience – The Development Dimension. United Nations. Available at: https://unctad.org/system/files/official-document/tdr2021_en.pdf

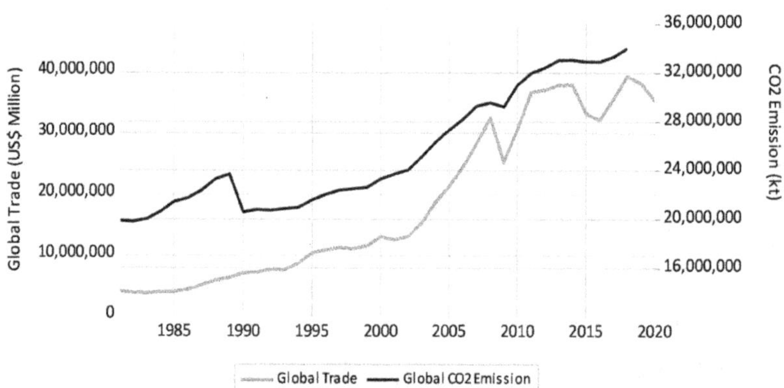

Figure 7.9 Trajectory of Global Merchandise Trade and Carbon Dioxide (CO2) Emissions. United Nations Conference on Trade and Development. (2021). Trade and Development Report 2021: From Recovery to Resilience – The Development Dimension. United Nations. Available at: https://unctad.org/system/files/official-document/tdr2021_en.pdf

from the global financial crisis, crude oil crisis and the coronavirus pandemic, respectively (see Figure 7.7).

It is important to note that countries strive to achieve a desirable level of economic growth capable of lifting a larger proportion of her populace out of poverty, ensure shared prosperity and improve the general welfare of its citizenry. One avenue to achieving economic growth has been through cross-border trade. However, though cross border trade is beneficial as it improves domestic welfare, raise income level, address domestic unemployment, increase tax revenue, bringing about technological advancement, it could have significant impact on the environment and bring about environmental degradation. A critical assessment of the trend movement of global trade and greenhouse gases emissions as depicted by Figure 7.8 and 7.9 revealed that greenhouse gases (GHG) emissions and carbon dioxide emissions seem to increase with rise in cross border trade. The graphical illustrations suggest some form of causal relationship between global trade and climate change. Empirically, there has been evidence that climate change induced by greenhouse gases (GHG) emissions is caused by increased international trade (Zhong, Guo and Zhang, 2021; Shi and Trigkas, 2021; Wang, Tang, Du, and Song, 2020; Li and Wang, 2019; Du, Yu, and Li, 2020).

Traditionally, there is theoretical dichotomy on the relationship between climate change and cross border trade. Empirical findings seem to support this divide as the relationship between climate change and cross border trade seems to be bidirectional. Recent analysis on cross-border trade reveals that, in the long run, global trade and its size relative to global income is driven by four factors such as:

i) Transportation and communication costs,
ii) Growth in income and changes in preferences,
iii) Sectoral comparative advantage in production of goods and services, and
iv) Trade policies and trade agreements.

A country's specialization and changes in specialization depends largely on the differences in these drivers among countries. Traditional trade theory argues that a country's specialization in particular goods is determined considerably by the availability of inputs used in manufacturing the product and access to necessary technologies.

Climate change tends to affect some of these elements, thereby altering cross border trade and specialization patterns through several mechanisms. Climate change in the form of rising sea levels or higher frequency of extreme weather condition will directly impact on trade as it will affect transportation and disrupt global distribution chains (GDC). Furthermore, the changes in factor inputs (labour, land, and capital) brought about by climate change will affect production structure and invariably trade specialization. The effect of climate change on trade could also be indirect, as sectors and regions are linked through inputs in production and exchange of manufactured goods and services. Thus, the consequences of climate change on trade can be direct and indirect. The direct impact of climate change on trade encompasses alteration to transport and distribution chains and changes to trade-relevant supply. The effect of these could be visible in terms of damages to trade infrastructure such as ports.

Extreme weather events can lead to opening up of new trade routes. The indirect impact is the changes it makes to production of goods and services through adjustment to factor inputs. The combination of these direct and indirect effects of climate change can alter the comparative advantage of a country, thus change the flow of global trade and pattern of trade. According to Hummels (2007), one of the reasons that embodied international trade and made it prominent in the recent decades, is the declining international transportation costs. Companies or firms have set up firms in other countries or engage in trade with other economies due to leverage of low transportation cost as it makes economic sense to do so. However, supply, transport and distribution chains may become susceptible to disruptions due to climate change. Ports and shipping routes may be temporarily shut due to extreme weather events induced by climate change. This could lead to delays, increased insurance costs, cause damages to trade infrastructures, raise trade cost, prices of import commodities and result in shift in trade patterns as companies try to adjust to alternate shipping route (Heininen, et al., 2015).

Apart from land-based transportation consequences of climate change in the form of faster degradation of bridges and road infrastructure, airborne transportation of merchandize goods may be affected by climate change. Operations of airports may be affected by increased storminess, raise sea level induced by climate change. Studies showed that under these extreme weather conditions, airlifting of merchandise goods may be difficult. Climate change may have negative effect on maritime shipping, which according to UNCTAD (2019) accounts for 80 percent of global trade. Rising sea levels and increased storms may affect speed of ships and cause ship operators to utilize additional safety measures and take alternative route which may increase arrival time, increase transportation costs and maintenance costs for ports and ships.

The indirect impact of climate change on cross border trade can manifest in the form of changes in crop yield, loss of capital and land necessitated by rising sea level, changes in energy demand for heating and cooling, disruption in fishing activities, capital damages from hurricanes, among others. The combination of these effects could significantly affect the gross domestic product of the global economy. The impact on energy demands may affect energy prices and propel changes in production in energy-intensive industrial sectors. The changes in crop yield resulting from frequent extreme heat and increase in temperature could pose serious challenges to food security, negatively affecting the macroeconomy.

Cross border trade has been a source of greenhouse gases emissions and propeller of climate change. This argument flows from the idea that cross border trade and economic growth go hand in hand and industrialization has been the major driver of economic growth. Economic and environmental theories indicate that countries that prioritize economic growth will be a victim of environmental degradation from associated industrialization, natural resource depletion and pollution. Opposing theories of sustainable development counters this by suggesting that economic growth generates income and wealth which facilitates investment in environmental infrastructure and help mitigate environmental degradation possible through access to advanced technology and formulation of environmentally friendly policies and practices.

The relationship between cross border trade and climate has been empirically investigated and rapid growth in international trade has been found to increase carbon emissions. Zhong, Guo and Zhang (2021) investigated the impact of international trade on carbon emission in 39 major economies and found that development in international trade is responsible for 5 percent increase in carbon emissions. Shi and Trigkas (2021) who conducted a study in the Belt and Road region, discovered that a non-linear relationship exists between international trade and carbon emissions. On the other hand, Wang, Tang, Du, and Song (2020) discovered that the increased energy consumption in China was instigated by the expansion in international trade. Du, Yu, and Li (2020) who studied 116 countries from 1986 to 2014, found that international trade can reduce carbon emissions and increase income. Li and Wang (2019) also found that increase in international trade can lead to reduction in carbon emissions in China.

Cross border trade may affect climate change through its direct impact on carbon emissions and other greenhouse gases (GHG). Trade affects greenhouse gases (GHG) emissions in multiple ways. The effect of cross border trade on climate change extends beyond greenhouse gases from production and international transportation. The production of goods and services for trade and their consumption generates greenhouse gases emissions as the production processes involves the use of equipment and machinery that are large emitters of greenhouse gases such as carbon dioxide, methane, and nitrous oxide. Figure 7.8 reveals that evolution in global trade and greenhouse gases emissions follow similar pattern as greenhouse gases emissions increases with expansion in international trade.

The transportation of merchandise trade through ships and airplanes directly and indirectly contribute to greenhouse gases (GHG) emissions. According to ITF (2021), over half of international trade-related carbon emissions are from maritime shipping. The amount of carbon emissions embedded in trade is determined by the following factors.

i. Size of the economy, as large economies is the largest exporters and importers of greenhouse gases emissions. Size of population also drives GHG emissions embedded in trade.

ii. Sectoral composition of trade flows. Economies specializing in GHG-intensive sectors like petroleum and coal products, chemicals and pharmaceuticals, and transportation will contribute more to greenhouse gases emissions through trade.

CLIMATE CHANGE AND BUSINESS SURVIVAL

There are growing concerns on the implications of climate change on business survival. The spill-over effects of climate change on small-scale business tend to be more severe considering their poor adaption strategies. The IPCC (2014) and Bueno and Soto (2017) 0bserve that most climate change predictions have shown that small-scale producers will be the most affected by variability in climatic conditions and changing weather events due to their poor adaptive capacity. Harvey *et al.* (2018) posit that the changing pattern in climate unpredictability is detrimental to smallholder farmers and can undermine global progress toward poverty, food security, and sustainable development by limiting the level of progress on the SDGs. This makes the smallholder farmers highly vulnerable to climate change impacts because they are mostly dependent on rain-fed agriculture, cultivate marginal lands coupled with limited access to technical and financial support that could help them venture into more climate-resilient agriculture. Adapting to climate change will involve taking the right actions to reduce the harmful impacts of climate change on them. This can be done by making appropriate adjustments and changes.

Furthermore, it has acknowledged that climate change will intensify existing problems of business while creating new combinations of risks which pose a threat to the survival of businesses worldwide. The situation will be made worst due to widespread poverty, dependence on rain fed agriculture, land tenure systems, limitations in capital and technology accessibility, inadequate infrastructure, and inadequate research and extension. Hence, adaptation will be the key defensive measure. Adaptation to climate change will involve changes in agricultural management practices as a response to changes in climate conditions (Akinnagbe & Irohibe, 2014). Smallholder farmers and other businesses in many developing countries are confronted with an urgent need to establish resilience to the impacts of climate change. However, the Food and Agriculture Organization (FAO, 2021) observe that it will be a difficult task eradicating global poverty and hunger without developing resilience to climate change in smallholder agriculture through adaptation and sustainable land, water, fisheries, and forestry management practices.

O'Brien (2008) opines that climate change has very strong relationship with the recent financial crisis and businesses, manifesting in the form of new business opportunities, among others. He further explained that this may cause insurers to face high premiums or a complete elimination of private coverage options. The implications of climate change and changing weather events tend to trigger new ways of doing business, new legislation on the part of government, new policy framework, and new business challenges alongside new methods of adaptation for sustainability. This threatens business survival with increasing cost and dampening effects on profitable growth. Again, changing climatic conditions can adversely affect the purchasing power of consumers and, in so doing distort consumption smoothening to the detriment of some businesses. This creates difficulties to the survival businesses.

More frequent extreme weather events such as heat waves, droughts and heavy precipitation are detrimental to survival business activities. Aside from frequent extreme weather events, gradual climatic processes such as sea level rise, rising temperatures and changing rainfall patterns equally undermine profitable growth of businesses by increasing the risks and uncertainty of business survival. As the weather events continue to evolve with changing climatic events, it is imperative for the development of measures to address the potential risks and uncertainty these pose on business operations and survival.

CONCLUSION

The implications of the changing dimensions of climatic processes and weather events on cross-border trade and business survival has continued to attract the attention of policymakers and other stakeholders in the environment. Small businesses have been adjudged as one of the most vulnerable economic agents to current climate variability which also has very strong impacts on the global economy. This vulnerability is exacerbated by other challenges like endemic poverty, limited access to appropriate infrastructure and the accompanying technology as well as limited access to capital and markets. Thus, the intended and desired environment-specific mitigation and adaptation strategies should be developed and implemented; especially in poorer economies in Africa that are very vulnerable to climate change processes and changing weather events due to poor adaptive capacity. These will create opportunity for these poorer economies to benefit more from international trade and cooperation.

Furthermore, to ameliorate the negative effects of climate change on cross-border trade, there should be clear, transparent, and reliable framework for international cooperation on climate change adaptions which would have a sound economic basis for proper cost-benefit analysis. The international cooperation on climate change adaptions should promote climate-friendly investments with low carbon technologies and cleaner energy sources to allow countries optimize the benefits associated with cross-border trade. In addition, adequate insurance schemes should be provided for businesses in regions with poor adaptive capacity to enable them build resilience against the effects of climate change and improve their potentials for survival. Governments should synergize with the private sector and other stakeholders in the environment to ensure that policy formulations related to climate change adaptations and interventions allocate more resources toward the most vulnerable business groups based on the recognition that climate change will affect businesses differently in view of their specific region, environment, knowledge regarding climate change and economic conditions, among others.

REFERENCES

Ajibade, L., & Eche, J. (2017). Indigenous knowledge for climate change adaptation in Nigeria. *Indigenous Knowledge Systems and Climate Change Management in Africa Working Paper No.* 219.

Ding, D. K., & Beh, S. E. (2022). Climate Change and Sustainability in ASEAN Countries. *Sustainability*, *14*(2), 999.

Du, K., Yu, Y., & Li, J. (2020). Does international trade promote CO2 emission performance? An empirical analysis based on a partially linear functional-coefficient panel data model. *Energy Economics*, *92*, 104983.

Enete, A. A., &Amusa, T. A. (2010). Challenges of agricultural adaptation to climate change in Nigeria: A synthesis from the literature. *Field Actions Science Reports. The Journal of Field Actions*, *4*.

German Advisory Council on Global Change (2003) Climate Protection Strategies for the 21st Century: Kyoto and beyond. Special Report. Berlin, Germany, WBGU: 1.

Heininen, L., Exner-Pirot, H., & Plouffe, J. (2015). *Artic Yearbook 2015*. Northern Research Forum.

Huang, J. & Wang, Y. (2014).Farmer's Adaptation to Extreme Weather Events through Farm Management and Its. *Land Economics*, *90*(1), 26-43.

Hummels, D. (2007). Transportation costs and international trade in the second era of globalization. *Journal of Economic perspectives*, *21*(3), 131-154.

IPCC (2007). Climate Change 2007: Synthesis Report. Contribution of Working Groups I, II and III to the Fourth Assessment Report of the Intergovernmental Panel on Climate Change. *Technical Report Intergovernmental Panel on Climate Change.*

IPCC (2014). *Climate change 2014: Synthesis report. Contribution of Working Groups I, II and III to the Fifth Assessment Report on the Intergovernmental Panel on Climate Change*. Core writing team, R. K. Pachauri, and L.A. Meyer. Geneva: Intergovernmental Panel on Climate Change, 151 pp. Available online at: http://www.ipcc.ch/pdf/assessmentreport/ar5/syr/SYR_AR5_FINAL_full_wcover.pdf

Kandu, P. (2017). Papua New Guinea. Impacts of climate variations on local fisheries and aquaculture resources in PNG. *Ecological Risk Assessment of Impacts of Climate Change on Fisheries and Aquaculture Resources, ed EJ Ramos (Peru: APEC Ocean and Fisheries Working Group)*, *45*, 49.

Li, S., & Wang, S. (2019). Examining the effects of socioeconomic development on China's carbon productivity: A panel data analysis. *Science of the Total Environment*, *659*, 681-690.

Raza, A., Sui, H., Jermsittiparsert, K., Żukiewicz-Sobczak, W., & Sobczak, P. (2021). Trade liberalization and environmental performance index: Mediation role of climate change performance and greenfield investment. *Sustainability*, *13*(17), 9734.

Shi, Z., & Trigkas, V. (2021). Reconstructing Geography, Power and Politics in the Belt and Road Initiative, *Investment Research, 6,* 67-92.

UNCTAD. (2019). World Investment Report 2019: Special economic zones. New York and Geneva: United Nations, pp. 126-206.

Wang, S., Tang, Y., Du, Z., & Song, M. (2020). Export trade, embodied carbon emissions, and environmental pollution: An empirical analysis of China's high-and new-technology industries. *Journal of Environmental Management*, *276*, 111371.

Zhong, Z., Guo, Z., & Zhang, J. (2021). Does the participation in global value chains promote interregional carbon emissions transferring via trade? Evidence from 39 major economies. *Technological Forecasting and Social Change*, *169*, 120806.

Chapter 8

The Corona Virus Pandemic

Health and Economic Implications

By Priscilla Okunji, Dana Mashal, Caron Strong, and Devora Winkfield

INTRODUCTION

1918 was a difficult year for the worldwide population for many reasons. Primarily, the world was in chaos due to World War I. The world had been in this war for roughly four years, and at the time, there was no end in sight. Unknowingly, another war of human proportions was waging quietly in 1918, the H1N1 influenza pandemic, also colloquially known as the Spanish Flu or Great Influenza (Center for Disease Control [CDC], 2018). The CDC further pronounced, the Spanish Flu was the most severe influenza outbreak in the 20th Century and that it was among one of the most harmful and tragic pandemics in history.

Contrary to its name, the Spanish Flu did not in fact originate in Spain. The first public news of the disease was reported out of Madrid Spain, in May 1918, hence the name Spanish flu was coined (Liang et al., 2021). More accurately, Martini et al. (2019), stated that the first documented case of the Spanish Flu was in the United States (US) in the quiet and unassuming state of Kansas. During World War I, several small towns in America established military camps, one of these camps in Kansas saw the emergence of influenza. Its rapid spread led to nearly 1,100 soldiers being afflicted with the disease in only three weeks. At the time, information was not freely released to the public in order to avoid fear and hysteria. As a result, this created challenges in accurately mapping the origins of the virus and its spread to the world.

The Spanish Flu, now known to be avian in nature, came in three waves: the first began in March of 1918, the second in mid-August 1918, and the third in January 1919 (CDC, 2018). Reports by Martini et al. (2019), detailed that the first wave of the flu spread through Europe, specifically France, Great Britain, Italy, and Spain. In Europe as well, the primary vectors were military personnel. By May 1918, the flu had spread to North Africa and then to India. A month later in June, the flu took hold of China, and then quickly spread to Australia in July.

After a brief lull, the second wave of flu sprung back in the late summer of 1918. It spread from England to both Africa and North America and as with the first wave was assumed to have been moving with the armies. Lasting almost six weeks, the second wave spread through the rest of Europe reaching Eurasia countries such as Russia, India, and China. Meanwhile, in the US, scientists were optimistic that the virus would have dissipated by November 1918, even though it had persisted in Europe and Asia. The second wave was so strong and virulent; it was recorded to be the deadliest variant of the pandemic. The virus affected many young healthy individuals who were burdened with the highest mortality rates (CDC, 2018). Schwartz (2018) details the US Public Health Service swift and robust activation plans during the second wave to curb the virus' spread. They pushed public hygiene campaigns to control the spread. Even more aggressive practices such as school closures, restrictions on gatherings, isolation and quarantines were instituted. The second wave was so ruthless in America; the CDC (2018) chronicles its effects and describes its grip on the public health system. They conceded that the US health system was overtaxed with the war and the virus. There was a shortage of professional nurses and a call for volunteers and accelerated nursing programs. Morgues were overwhelmed and makeshift hospitals created as communities were decimated. The economy struggled further due to the war and then the virus' toll on workers. Some states initiated reportable status for flu and many initiated isolation and quarantine policies.

According to Martini et al. (2019), the third and final surge in the flu virus blasted the world in January 1919. The third wave is said to have hit Australia first, then New York, and Paris. This wave held its grip on most of the world until May 1919, thereafter the pandemic ended in most of the world. However, a small pocket of the virus persisted in Japan until 1920. In total, the pandemic's mortality rate was more than 50 million people and had infected more than 500 million people. It was recorded that of the total death toll, 100,000 of those were military troops, the unassuming host and super spreader of the virus worldwide.

In an attempt to combat the flu scourge, scientist feverishly started the development of a vaccine in 1918 (Schwartz, 2018). A historical review of the annuals by Schwartz (2018) discovered that at the time, it was believed that the flu was caused by a bacterium due to research that was done by physician and renowned scientist Richard Pfeiffer in the 1890s. Bacillus influenzae (now Haemophilus influenzae) was found among those with the

Spanish Flu. Vaccine research targeted the bacteria theory at the time which resulted in poor outcomes and disappointment among citizens and healthcare professionals alike.

Unknowing health professionals felt they found a panacea for the virus and began to spread word that a "revolutionary" vaccine was available that could protect the public from the Spanish Flu. Subsequently, the public had a period of confusion, because while their physicians were recommending the vaccine, the Surgeon General and other public health officials were trying to lower the public expectation for the new vaccine. Many local and Federal health agencies vehemently objected to the claims of the vaccines and publicly rejected their claims. Unfortunately, the vaccines that were developed in the early time of the pandemic did not have much recorded effect and the warning from the healthcare officials were accurate. The field of virology was in its infancy at the time, and many scientists held the belief that the Spanish Flu was in fact a virus. It wasn't until the 1930s that headway was made in the development of a flu vaccine after an explosion in the science of virology. In 1933 and 1940 it was found that there were two main strains of the influenza virus, influenza A and B, respectively. By 1936, two influenza A vaccines were developed. This led to large scale use of the influenza vaccine by the end of World War II and the second half of the 20th century which had a tremendous impact on the spread of influenza.

THE COVID-19 PANDEMIC

The first case of COVID was reported in Wuhan City, China in December 2019 (CDC, 2022). The CDC chronicles the rapid spread of the virus from a Wuhan City seafood market to Thailand and from there spread to other countries around the world including Japan, US, and Italy. The first case in the US dates back to January 2022, from Washington state. Due to the virus' significant spread, the World Health Organization (WHO) officially declared the world to be in a pandemic on March 11, 2020, just 3 months after the first identified case. Though they had issued a warning in January 2020 after local Chinese health departments issued concerns the end of December. It is noteworthy that the first case of COVID was not directly link to the seafood market. Every region of the world was infected and paralyzed by the virus only months after it was discovered. Liu et al. (2020) asserts that COVID-19 is the fifth pandemic the world has seen since the 1918 flu pandemic. Other outbreaks of influenza and other viruses such as HIV, Ebola, and coronaviruses (SARS and MERS), have also captured the attention of the world. COVID-19, a novel coronavirus, has held its grip on the world for the greater part of two years. Similar to the Spanish Flu in its animal origins, COVID-19 is suspected to have originated from bats, due to a 96% identical match of the DNA make-up of the virus to the Bat-CoV RaTG13 (the equivalent of COVID-19 virus in animals). Apparently, COVID-19 shares many similarities to the Spanish Flu in its virulence, spread, paralysis of the world, mortality, morbidity, scientific and healthcare response, and crippling economic outcomes.

Although the virus was first found in China, it has continuously mutated, and multiple variants have originated. While this is not uncommon with viruses, it has caused complications in treatment, prevention, and vaccine production (CDC, 2021). From Wuhan City to other areas in China, the virus quickly spread external to China. After arriving in the US, especially with the rate of international tourism caused the virus to spread like wildfire. It was due to the rapid spread and different outbreaks and peaks that new variants emerged (CDC, 2021).

McLean et al. (2022), outlines the chronology of the virus' evolution. The original virus that began towards the end of 2019-beginning of 2020 evolved into the Alpha variant in September 2020 in Great Britain. In December 2020, strong hopes for controlling the virus were realized with the development of a vaccine. While Alpha was spreading from Great Britain and Europe, South Africa was seeing the Beta variant emerge in October 2020. The Gamma variant was quickly identified in Brazil in December 2020. After the decline of these three variants, India saw the Delta variant in late 2020. The current Omicron variant was first seen late November of 2021 in South Africa. It's BA sub-variants are still dominant around the world (Katella, 2022). As a result of these variants and ease of transmissibility COVID cases continue to soar globally. As of August 2022, the Kaiser Foundation reports the total number COVID cases around 599 million, with a total of 6.4 million deaths around the world (KFF, 2022).

In December 2020, vaccine research had produced hopes for controlling the virus (McLean et al., 2022). Swift and rigorous development and trials resulted in highly efficacious vaccines. The vaccines were administered in one or two dose series. Shortly after the emergence of the first vaccine, a booster shot was recommended five months after the completion of the original vaccine. It is now also recommended that a second booster to be administered roughly four months after the first booster (CDC, 2021). Mathieu, et al. (2020) state that 67.7% of the world population has received at least one COVID vaccine. There continues to be widespread hesitancy in receiving the COVID-19 vaccine. An article by Dhama et al. (2021) studied the various reasons why citizens around the world were choosing not to get the vaccine. Some of the main reasons that they found were safety and side-effect concerns, lack of understanding on how vaccines are developed, fear of shortcuts being made due to speed of development of the vaccine, and ultimately mistrust and uncertainty in medical companies, healthcare professionals and government officials.

THE PANDEMIC PREVALENCE AND MORTALITY RATES IN THE US AND THE WORLD

COVID-19 quickly reached pandemic proportions due to its transmissibility and virulence. The CDC (2022) chronicles the COVID-19 outbreak from the first official case of the disease in China in December 2019. In 3 months, the virus was found in most countries and had significant casualties, making it's mark as one of the most widespread and deadliest diseases the world had ever seen. The rapid changing characteristic of the COVID-19 virus made it even more difficult to treat. Many variants have and are expected to continue to emerge as a natural characteristic of viruses (Liu et al., 2020). The original virus that began during the end of 2019 -beginning of 2020 evolved into the Alpha variant in November 2020 in Great Britain. While Alpha was spreading from Great Britain and Europe, South Africa was seeing the Beta variant at the same time. After the decline of the Alpha and Beta variants, India saw the Delta variant in the late 2020 as well, which overtook the numbers of the other two variants. The current Omicron variant and its sub-lineages was first seen late November of 2021 and remains the dominant variant around the world (Katella, 2022).

The prevalence of COVID-19 in the US remains a concern. The CDC (2022a) archives cases from April 2020 until today, there has been a total number of over 94 million cases reported in the US. Of those reported cases, 5,264,088 individuals were admitted to the hospital due to COVID-19 related complications. There have been over a million Americans who have died from the virus, with states such as California, Texas, Florida, and New York reporting the highest losses. Of note, over 224 million people (about 67.5% of the total population) are fully vaccinated. With 108 million (48.5%) have received the first booster vaccine.

Jones (2022) explains that the death toll and impact varied based on the variant at large. From the beginning of the pandemic, March 2020 through June 2020, the US saw roughly 125,000 deaths in that three-month period. The summer of 2020 until the wintertime, another 80,000 Americans died from the disease. It is considered that the fall-winter of 2020 into April of 2021 was the deadliest wave of the pandemic, with more than 370,000 recorded deaths in the US. Jones (2022) adds that there was a period of reprieve with the rise in numbers of citizens receiving the vaccine, resulting in the US seeing a decline in the death toll. Combining the fourth and fifth waves of the pandemic, the death of more than 300,000 Americans occurred.

Globally the numbers are equally devastating. The WHO (2022), reports that there are over 603 million confirmed cases of COVID-19 around the world. Of those cases, unfortunately 6,484,136 people have died of the disease. Reassuringly, there is close to 5 billion people worldwide who received at least 2 doses of the vaccine. A comparison of countries found that the US was leading in both confirmed cases, close to a million, and deaths, over a million from COVID. Among deaths, India and Brazil are second and third with 684 million and 528 million, respectively. India and Brazil also have the second and third leading cases behind the US. The world saw a large spike in case numbers from around December 27, 2021, through April 11, 2022, due to the Omicron variant and its sub-lineages. There have been multiple spikes in deaths throughout the entire pandemic. The most noteworthy was winter 2020 through February 2021, again March 2021-June 2021, and again January 2022-March 2022.

THE PANDEMIC IMPACT WITH THE DEVELOPMENT OF VACCINES USING THE MRNA TECHNOLOGIES

The pandemic brought about two trailblazing processes to the world and science, the mRNA vaccine and the Food and Drug Administration (FDA) power to grant emergency use authorizations (EUAs) (Iwry, 2021). In December 2020, the mRNA vaccine began to emerge after rigorous testing and trials. Though the science had been around for decades, the pandemic thrust it into further development and testing. Both Pfizer/BioNTech and Moderna vaccines are administered in two dose series. Shortly after the emergence of the first two vaccine doses, a booster shot was recommended five months after the completion of the original rounds. It is now also recommended that a second booster be administered roughly four months after the first booster for those at high risk. More recently, an Omicron specific vaccine has been available to combat the spread of the variant (McLean, et al, 2022).

A main concern that society had with the creation of the COVID-19 vaccines was the mRNA vaccine technology (Anand & Stahel, 2021). The CDC (2022b) discusses that the mRNA, or messenger RNA, tells the body to create specific proteins called spike proteins. With the production of these proteins, the body's immune system can now produce the specific antibodies that can then fights off the actual virus were the person exposed or infected. This mRNA technology was first discovered by Drew Weissman, MD, PhD, and Katalin Karikó, PhD in 2005, which paved the way for Pfizer/BioNTech and Moderna covid 19 vaccines (Gardner, 2021).

Another concern that emerged was the speed at which the vaccine was developed (Anand & Stahel, 2021). Scientists had been researching and working on a coronavirus vaccine during the original SARS (Severe Acute Respiratory Syndrome) and MERS (Middle East Respiratory Syndrome) virus outbreaks in the early 2000. Contrary to public belief, there were no short cuts taken and no steps in the system created by the CDC or FDA skipped. Time and resources were primarily focused on vaccine development with the pandemic from governmental and private agencies. Scientists were able to expedite the process with technology instead of using the typical linear model of vaccine creation, they moved through the steps as if the previous one had been completely successful (BJC, 2021; CDC, 2022c). Through the clinical trials it was revealed that both vaccines (Pfizer/BioNTech and Moderna) had an efficacy rate of 94-95% after two doses (Anand & Stahel, 2021). After the clinical trials results were reported to the CDC and FDA, and EUA was granted for the use of both vaccines. An EUA was last granted during the H1N1 influenza season and in that situation, it had not been for a vaccine but for medical equipment (The Petrie-Flom Center Staff, 2021).

Anand & Stahel (2021), summarized the drug trials for the mRNA vaccines. The Pfizer/BioNTech trial had a total of 43,548 participants and were split into two different groups, those who received the vaccine and those who received the placebo. During the trial, a total of 170 people contracted COVID-19 with only 8 being among those who were vaccinated. Furthermore, only one of those 8 was considered a severe case. In the Moderna trial, there was a total of 30,420 participants, who were also split in half. There was a total of 196 participants who contracted the disease with only 11 of those being fully vaccinated.

With the success of the mRNA vaccine, there is hope and visions for future use. It is noted by scientists that mRNA vaccines are easier to create than other vaccines and that they may be more effective than other types (Gardner, 2021). Currently there is research in applying mRNA vaccines to other infectious diseases, specifically sexually transmitted diseases, and influenza. Gardner (2021) highlighted Penn Medicine's research on an mRNA vaccine to fight the herpes simplex virus 2 (HSV-2). More noteworthy, there has already been animal trials with mice that have shown that complete immunity can be achieved with this technology for those with HSV infections. The benefit of an mRNA vaccine is that it can adapt on its own to the different strains of HSV-2. mRNA technology for influenza vaccines is still being investigated but appears promising.

VACCINE APATHY, HESITANCY AND POLITICAL MANDATES

There has been a widespread hesitancy in receiving the COVID-19 vaccine. This hesitancy has been described as a major threat that affects global recovery from the pandemic. The hesitancy halts any attempts and efforts towards eradicating COVID-19 and the efforts to accomplish herd immunity. Those who are not vaccinated can be carriers of the COVID-19 disease unbeknownst to them and continue the spread of the virus to others (Dhama et al, 2021). Since persons can be infected with COVID-19 and show no symptoms initially or throughout their time of infection, they serve as super spreaders of the disease to others. It is because of those individuals that there remains a continuous spike in numbers and new mutations. To accomplish herd immunity, there must be 60-90% of the total population who are vaccinated, however, due to many of the world's population choosing not to get vaccinated despite having access to them, many countries are now dealing with an increase in incidence of the virus as well as mutations and variants of the virus (Dhama et al, 2021). Some of the main reasons that people hesitate to be vaccine include safety and side-effect concerns, lack of understanding on how vaccines are developed, fear of shortcuts being made due to speed of development of the vaccine, and ultimately mistrust and uncertainty in medical companies, healthcare professionals and government officials (Dhama et al. 2021).

Markowitz & Rough (2022) reflects on early governmental mandates to control COVID-19. In the US, mandates were from Federal and State levels. Many, if not all, government and health officials began mandates and recommendations for schools, businesses, and public space. Understanding the infectivity and transmissibility of the virus, public health practices to curtail the virus saw many closures, capacity limits, and mask rules for the public. When new variants or outbreaks occurred, public mandates become stricter, and when the incidence decreased the mandates tend to be lifted or relaxed. Though many mandates were state based and varied among states, the Federal government implemented many restrictions and policies. For example, President Biden issued an executive order that required all individuals in the healthcare field to be vaccinated. Another executive order that was issued due to the pandemic was mask wearing on any form of public transportation.

THE ECONOMIC TOLL IN THE US AND THE GLOBAL WORLD

COVID-19 did not only affect the world's population health, but it also took a toll on the world's economy. At the beginning of the pandemic, more than "tens of millions of people" were unemployed due to lockdowns (Center on Budget and Policy Priorities [CBPP], 2022). By the end of 2021, even though the US began to see a rebound from the drastic effects of the pandemic, 20 million households still stated that they were unable to adequately feed their family. In addition, 10 million households even stated that they remained unable to pay their rent. Currently, with society starting to return to the new normal there still remains "3 million fewer people who are employed because of the COVID-19 pandemic" (CBPP, 2022).

The discrepancy in household income due to the pandemic was discussed by Kochhar & Sechopoulos (2022). They emphasized that most households that were economically affected by the pandemic were those who were of middle to lower income levels. Between 2019 and 2020, it was observed that lower-income household had a median income decrease of 3% while middle-income households saw a median income decline of 2.1%. However, this contrasted with upper-income households, who did not experience an income decrease.

Globally, the International Monetary Fund (IMF) estimated that the median global GDP decreased by 3.9% from 2019 to 2020 (Kates et al., 2022). This estimation makes it the worst economic decline since that of the Great Depression. It was predicted that with the distribution of the vaccines, the world's economy would bounce back from its decline. Unfortunately, these predictions were overly optimistic; the IMF reports that the economy remains way below pre-pandemic estimations (Kates et al., 2022).

THE PANDEMIC ON HUMAN MIGRATION, NEW ORDER, EMERGING TRENDS, AND GLOBAL REALITIES.

The Role of the Pandemic in Human Migration

This final segment will investigate the role of the pandemic in human migration, new order, emerging trends, and global realities.

During the COVID-19 pandemic, many nations around the world enforced quarantines, entry bans, and other travel restrictions for their citizens. By restricting travel, many countries saw a decline in the spread of the virus. Unfortunately, these restrictions were implemented after the virus had already spread to multiple countries in different regions of the world. From March 2020 to February 2022, there was a total of 122,823 movement restrictions around the world (Migration Data Portal, 2022). During that same time, there were 961 exceptions that were made by 208 countries. Due to the many restrictions and requirements for travel, the world saw a decrease in migration of about 2 million people from mid-2019 through mid-2020 (Migration Data Portal, 2022).

Migration data sources concluded that tourists were not the only groups affected by the travel mandates and restrictions, but international students were affected as well. These students were restricted by both their country of origin and their destination countries. Furthermore, due to the travel restrictions and border closures imposed by global governments, several migrants seeking asylum and opportunities were left stranded and unable to return to their home country. It has been noted that to date there are at least 3 million stranded migrants (Migration Data Portal, 2022). McClanahan (2022) showcases the rebirth of leisure travel. With the downward trend of positive cases, the requirements for traveling have begun to loosen. It has been noted that some countries that had previously stopped allowing tourists to enter their country, have now either lifted bans or currently planning to. The CDC, however, continues to recommend that those planning to travel internationally wait until they are fully vaccinated against COVID-19. While the pandemic did not completely halt tourism and travel, the world saw a significant decline. This decline not only damaged the world economy, but also disrupted people's vacations and social activities. There were many weddings and funerals where families were not able to come together to celebrate.

The New Order and Emerging Trends in E-learning and Hybrid-Virtual Instruction

COVID-19 saw the rapid burgeoning of working-studying from home. The pandemic forced schools to close across the world. Over 1.2 billion children were out of the formal classroom (Li & Lilani, 2020). As a result, education has changed dramatically, with the distinctive rise of e-learning, whereby teaching is undertaken remotely and on digital platforms. Li & Lilani (2020), proport that many studies have suggested that online learning has been shown to increase retention of information and take less commuting and preparing time. This form of learning was tested strongly during the pandemic and some forms may well be here to stay. But E-learning is not without its challenges, almost 97% of educators reported that they have noticed academic delays in their students with the year of virtual learning versus students from previous years (Dickler, 2021). Dickler (2021) goes on to add that 57% of teachers also stated that their students are behind by more than three months in their social-emotional development. Teachers from eight different countries reported that remote learning is a poor substitute for being back in the traditional classroom. The US and Japan gave distance learning the harshest scores overall, with most teachers ranking the effectiveness only slightly better than skipping school completely. While virtual learning helped with minimizing the spread of the virus, there were obstacles that arose. Some students struggled with attendance due to lack of access to reliable internet and/or technology and electronic resources (Li & Lilani, 2020).

The pandemic also redefined the traditional workplace. Many realized that they do not need to be tied to their office to successfully complete their work.

In fact, many of those who are currently working from home reported that they are now choosing to work from home when given the option to return to the office (Parker et al., 2022). Parker et al. (2022) asserts intriguing telework data. They stated that only 38% of people who are currently working from home are saying that this is due to their workplace being closed. Comparatively, at the start of the pandemic, 64% were working from home due to office closures and 36% said that they are working from home out of preference. Working from home is a new concept for many workers with jobs that can be done remotely, as much as 57% said they had never worked remotely. A positive outcome from telework was that 64% stated that they are better able to balance work with their personal life. Moreover, with teleworking, 44% stated that it is easier for them to get their work done and meet deadlines.

The Current China COVID-19 Prevalence and its Implication to a Second Omicron Variant Outbreak in 2022 in the US and the World.

Like two years ago when the pandemic first began, China is now seeing a large surge in COVID-19 cases. It has been reported that this surge began at the beginning of March 2022 and has been going strong ever since (Yeung, 2022). By March 6, experts and officials began warning the country that the outbreak was severe but that the country had the capability to control it. Many major cities were tightening their COVID restrictions as the country struggled to contain a nationwide outbreak stretching from Shanghai in the southeast to the Northern provinces. These restrictions and measures also include the closing of public schools for children and a return to online classes (Yeung, 2022).

According to the WHO (2022), China had been seeing only a few hundred cases each week from April 2020 until February 2022. The WHO reports that since February 2022 there have been a total of 968,473 cases of COVID-19. There has been a total of over 6 million cases of COVID19 from 2020 until today. China's safety measures have been successful in keeping their total death numbers below 30,000 for the entirety of the pandemic. China saw a surge of deaths at the beginning of the pandemic, from January through April 2020, then another small rise in deaths from June to July 2021. But, like case numbers, the majority of these deaths were recent from February through May 2022, totaling 9,573 deaths; this being roughly double the number of deaths from January to April 2020.

It has been reported that the variant that is the cause of this surge in China and now around the world is Omicron (Yeung, 2022). This variant has been hard to trace by researchers because symptoms are much milder than others as well as its incubation time is shorter. There are several subtypes of Omicron that are being seen; the original BA.1 and BA.2 and now BA.4 and BA.5 (CDC 2022d). According to the CDC, as of August 2022, 88% of

COVID-19 cases are due to the BA.5 lineage. The WHO (2022) reverberates the CDCs finding of Omicron BA.5 accounting for 90% of the COVID cases worldwide. With this new Omicron variant outbreak, lockdown measures have been reinstituted by China. Beijing has brought back mass testing, quarantines, and in-person schools have been shut down (Yeung, April 2022).

Global Realities Occasioned by the Pandemic as in Future Preparedness to Mitigate Mortality and Morbidity rates

While the world has not reached the point where the COVID-19 virus is eradicated, it can be said that there is a better understanding of the virus and that there is hope for the future to normalize society again. As previously mentioned, the virus, despite its added hardships, brought about some positive changes as well. Medical innovations such as vaccines, medications, rapid vs PCR testing, and natural immunity from contracting the virus, all helped scientists and governments to develop plans on how to handle outbreaks in the future (Usheroff, 2022).

The goal for the future, as stated by Dr. Soniya Gandhi from Cedars-Sinai, is for the virus to move from a pandemic to an endemic by both immunity through vaccines and naturally (Usheroff, 2022). She also describes another solution where, like the flu, there are seasonal surges or even more localized controlled outbreaks. Dr. Gandhi goes further to state that the worst option for the future is that the virus continues to mutate at a rate that scientists cannot predict or keep up with. Despite what government officials have decided, Dr. Gandhi believes that while we are still in a pandemic, society needs to remain vigilant by wearing masks while in heavily populated settings such as being indoors (i.e., offices, malls, school), traveling, and healthcare buildings. She also believes that if the virus moves down the path of having seasonal outbreaks that mask mandates will be needed during those times as well. In addition to masks and vaccines, Dr. Gandhi stated that their needs to be more healthcare technology and access to COVID-19 tests. Tests also need to become more sensitive to other variants as well as the public having access to quality at-home tests. By having good, reliable tests at home, those with exposures or suspicious of being positive can remain quarantined in their home and test themselves and avoid exposing others. Dr. Gandhi concludes with a call to action by the public to come to the realization and have some understanding and awareness that individuals' actions can and will impact the population. While the opinions of Dr. Gandhi are valid and appropriate, some infectious disease experts feel that there is a lack of understanding on what endemic means and how it affects society. Krista Conger (2022), reporter from of Stanford Medicine, spoke with Dr. Devaid Relman, professor of microbiology and immunology, and a former president of the Infectious

Diseases Society of America, on the topic of COVID-19 and what the future may hold. Dr. Relman stated that an endemic does not equate to harmless, and that the public needs to understand that tuberculosis and HIV are both extremely deadly and are classified as endemics. He further explains that "An endemic relationship speaks to a degree of adaptation that allows a stable co-existence" (Conger, 2022). Conger's (2022) article continues to explain what the future of testing may look like. The research team at Stanford has developed an antibody test in order to evaluate how immune the person is and if additional boosters are needed. They suggest that this should be used globally to help identify when a booster is needed, especially with new variants emerging and vaccine supplies not at full capacity.

CONCLUSION

Although both pandemics have several differences, many of the mistakes and lessons learned from 1918 flu pandemic were repeated in the current pandemic. On reflection, it seems that government officials struggled with handling the pandemic. Lessons and patterns from the Spanish Flu pandemic were identified early by the public health officials but political influences lagged to implement life saving measure to slow the spread and mortality. While in the end, isolation, hygiene, and masks were mandated worldwide, it took several months for it to go into effect. One noteworthy difference is that with the rise in medical technology, there was a faster call to action for a vaccine to combat the virus. Though both vaccines were seen as a panacea for the pandemics, the COVID vaccine held better scientific rigor, precision, and outcomes than early flu vaccines. Since its use, confirmed cases have been milder and less deadly. Unfortunately, even with proven benefits of the COVID vaccine, contemporary society remains suspicious and reluctant to vaccinate compared to a social view in the 1900s (Dharma, et al., 2021).

While the 1918 pandemic was the worst influenza outbreak in history, there have been others and with each one came a new vaccine to fight it. From then there is an annual flu vaccine to help prevent transmission of the current strain. With how the COVID-19 pandemic is evolving and new variants emerging every few months, scientists are hopeful that there will also be an annual COVID vaccine. Like the flu vaccine, the annual COVID vaccine will target dominant strains (Dharma, et al., 2021). It is imperative that governments worldwide along with national and global healthcare organizations improve and repair the public mistrust, eliminate global disparities, and invest in their public health system. Only then will the world see a decline in numbers.

While the COVID-19 pandemic was and still is a devastating feat that the world had to face with its severity and deadly nature, it did bring about some new, and much needed changes. New vaccine technologies were introduced with their potential for other diseases burgeoning. Though COVID-19 affected the world's economy negatively, it sparked ingenuity on how to work from home and created new age jobs, especially in the digital fields. While vaccine apathy and hesitancy were concepts that predated the pandemic, those who remain unvaccinated are now seeing the benefits of vaccination for preventing and decreasing the severity of COVID-19.

With the end in sight for the COVID-19 pandemic, some valuable lessons were learned. It is important to note that the troubles that the world experienced with this pandemic only helped to prepare the population for when the next pandemic occurs. Future experts and officials can take what was discovered in the past few years and apply the strategies that worked and avoid the ones that did not work. One important aspect that was learned with this pandemic is that airborne viruses are not impossible to manage and defeat if society has the proper equipment and resources. Long-term solutions are needed, such as increased air filtration in workplaces and schools, as well as the availability of high-quality, reusable masks.

COVID-19 and the pandemic is not without discrimination. Many of the efforts and supplies were unequally distributed. About half of the available vaccines have been distributed to high-income and western developed countries, while roughly 75% of people in Sub-Saharan Africa remain without access to the vaccines or even have their basic hygiene needs met (Conger, 2022). Even in countries of wealth, the majority of those affected by the disease and the economic fallout were minorities, immigrants, and those of lower socioeconomics.

On a more positive note, history was made during the pandemic with the development of the vaccine in only a year. The vaccine was not only developed in the year, but also approved as well, with impressive effectiveness close to 100%. It is predicted that a COVID-19 vaccine may be needed at least annually or biannually (Conger, 2022).

Finally, Weberg (2021) study describes the reasons certain situations transform over time when unthinkable events occur. The pandemic, despite its calamities, fits well into the unthinkable events category and has shined light on the US and global healthcare, research, economic and educational system, and the opportunities for improvement. There were opportunities to develop mRNA vaccines within a short period of time, spearheaded by a minority female scientist. We saw evolution in our educational system, as in hybrid and full online teaching and learning, which were not recognized by many institutions prior. A shift in global commerce and work was highlighted by the pandemic as well.

The US and global healthcare systems were reminded of the importance of public health policies and initiatives. The need to sure up the local, national, and global public health systems and communications is crucial. More specifically, the care of the most vulnerable and fragile was recognized with the pandemic. Geriatric care and lifestyle warrant further investigation, policies, and services. The elderly was a large group affected and lost by the pandemic. Social isolation, the process of death and dying, and the tenuous healthcare system were also vital lessons learned. We must create an infrastructure to protect and survive the next pandemic, which is guaranteed to come.

REFERENCE

Anand, P., & Stahel, V. P. (2021, May 18). *The Safety of Covid-19 mRNA Vaccines: A Review.* Patient Safety in Surgery. https://pssjournal.biomedcentral.com/articles/10.1186/s13037-021-00291-9#citeas

BJC. (2021, January 14). *Timeline of the covid-19 vaccine development.* BJC HealthCare. https://www.bjc.org/Coronavirus/COVID-19-Vaccines/COVID-19-Vaccines-Articles/ArtMID/6435/ArticleID/4617/Timeline-of-the-COVID-19-vaccine-development

CBPP. (2022, February 10). *Tracking the COVID-19 economy's effects on food, housing, and employment hardships.* Center on Budget and Policy Priorities. https://www.cbpp.org/research/poverty-and-inequality/tracking-the-covid-19-economys-effects-on-food-housing-and

Centers for Disease Control and Prevention. (2018, March 21). History of 1918 flu pandemic. https://www.cdc.gov/flu/pandemic-resources/1918-commemoration/1918-pandemic-history.htm

Centers for Disease Control and Prevention. (2022, August 22). COVID data tracker weekly review. Centers for Disease Control and Prevention. https://www.cdc.gov/coronavirus/2019-ncov/covid-data/covidview/index.html

Centers for Disease Control and Prevention. (2021, September 15). Science Brief: COVID-19 vaccines and vaccination. https://www.cdc.gov/coronavirus/2019-ncov/science/science-briefs/fully-vaccinated-people.html

Centers for Disease Control and Prevention. (2022, September). *Covid Data Tracker Weekly Review.* Centers for Disease Control and Prevention. https://covid.cdc.gov/covid-data-tracker

Centers for Disease Control and Prevention. (2022, September). An overview of COVID-19 vaccines. Centers for Disease Control and Prevention. https://www.cdc.gov/coronavirus/2019-ncov/vaccines/different-vaccines/overview-COVID-19-vaccines.html

Centers for Disease Control and Prevention. (2022, September). Developing COVID-19 vaccines. Centers for Disease Control and Prevention. https://www.cdc.gov/coronavirus/2019-ncov/vaccines/distributing/steps-ensure-safety.html

Conger, K. (2022, March 11). *Two years into the pandemic, medical scientists consider what the future may hold.* News Center. https://med.stanford.edu/news/all-news/2022/03/scientists-pandemic-future.html

Dhama, K., Sharun, K., Tiwari, R., Dhawan, M., Emran, T. B., Rabaan, A. A., & Alhumaid, S. (2021). Covid-19 vaccine hesitancy – reasons and solutions to achieve a successful global vaccination campaign to tackle the ongoing pandemic. Human Vaccines & Immunotherapeutics, 17(10), 3495–3499. https://doi.org/10.1080/21645515.2021.1926183 KFF (2022, May 13). Global COVID-19

Dickler, J. (2021, October 12). *Virtual School resulted in 'significant' academic learning loss, study finds*. CNBC. https://www.cnbc.com/2021/03/30/learning-loss-from-virtual-school-due-to-covid-is-significant-.html

Gardner, A. (2021, April 21). *How mRNA Vaccines Could Treat Other Infectious Diseases Beyond COVID-19*. Penn Medicine News. https://www.pennmedicine.org/news/news-blog/2021/april/how-mrna-vaccines-could-prevent-or-eliminate-infectious-diseases-beyond-covid19

Iwry, J. (2021, January 28). *From 9/11 to covid-19: A brief history of FDA emergency use authorization*. Bill of Health. https://blog.petrieflom.law.harvard.edu/2021/01/28/fda-emergency-use-authorization-history/

Jones, B. (2022, March 3). *The changing political geography of covid-19 over the last two years*. Pew Research Center - U.S. Politics & Policy. https://www.pewresearch.org/politics/2022/03/03/the-changing-political-geography-of-covid-19-over-the-last-two-years/

KFF (2022, May 13). Global COVID-19 Tracker - Updated as of May 13 https://www.kff.org/coronavirus-covid-19/issue-brief/global-covid-19-tracker/

Katella, K. (2022, March 30). Omicron, Delta, Alpha, and more: What to know about the coronavirus variants. Yale Medicine. https://www.yalemedicine.org/news/covid-19-variants-of-concern-omicron

Li, C., & Lilani, F. (2020, April 29). *The COVID-19 pandemic has changed education forever. this is how*. World Economic Forum. https://www.weforum.org/agenda/2020/04/coronavirus-education-global-covid19-online-digital-learning/

Liang, S. T., Liang, L. T., & Rosen, J. M. (2021). Covid-19: A comparison to the 1918 influenza and how we can defeat it. Postgraduate Medical Journal, 97(1147), 273–274. https://doi.org/10.1136/postgradmedj-2020-139070

Liu, Y. C., Kuo, R. L., & Shih, S. R. (2020). COVID-19: The first documented coronavirus pandemic in history. Biomedical journal, 43(4), 328–333. https://doi.org/10.1016/j.bj.2020.04.007

Martini, M., Gazzaniga, V., Bragazzi, N. L., & Barberis, I. (2019). The Spanish Influenza Pandemic: a lesson from history 100 years after 1918. Journal of preventive medicine and hygiene, 60(1), E64–E67. https://doi.org/10.15167/2421-4248/jpmh2019.60.1.1205

Mcclanahan, P. (2022, July 5). *I'm a U.S. citizen. where in the world can I go?* The New York Times. Retrieved from https://www.nytimes.com/article/coronavirus-travel-restrictions.html

McLean, K. J., Lee, B., Moore, P., Schulz, T. F., Muik, A., Sahin, U., Türeci, Ö., & Pather, S. (2022). The Impact of Evolving SARS-CoV-2 Mutations and Variants on COVID-19 Vaccines. mBio, 13(2). https://doi.org/10.1128/mbio.02979-21

Migration data relevant for the COVID-19 pandemic. Migration data portal. (2022, March 18). https://www.migrationdataportal.org/themes/migration-data-relevant-covid-19-pandemic

Oum, S., Kates, J., & Wexler, A. (2022, February 4). *Economic impact of covid-19 on PEPFAR countries*. Global Health Policy. https://www.kff.org/global-health-policy/issue-brief/economic-impact-of-covid-19-on-pepfar-countries/#:~:text=The%20toll%20the%20COVID%2D,since%20the%20Great%20Depression.

Parker, K., Horowitz, J. M., & Minkin, R. (2022, March 23). *Covid-19 pandemic continues to reshape work in America*. Pew Research Center's Social & Demographic Trends Project. https://www.pewresearch.org/social-trends/2022/02/16/covid-19-pandemic-continues-to-reshape-work-in-america/

Ritchie, H., Mathieu, E., Rodés-Guirao, L., Appel, C., Giattino, C., Ortiz-Ospina, E., Hasell, J., Macdonald,B., Beltekian, D., & and Roser, M.(2020). Coronavirus Pandemic (COVID-19). https://ourworldindata.org/coronavirus

Rough, J., & Markowitz, A. (2022, April 29). *State by state coronavirus-related restrictions*. Politics & Society: Government & Elections. https://www.aarp.org/politics-society/government-elections/info-2020/coronavirus-state-restrictions.html

Schwartz J. L. (2018). The Spanish Flu, Epidemics, and the Turn to Biomedical Responses. American journal of public health, 108(11), 1455–1458. https://doi.org/10.2105/AJPH.2018.304581

Senior, R. (2022, January 19). Similarities between Spanish flu and the COVID-19 pandemic. American Nurse. https://www.myamericannurse.com/similarities-between-spanish-flu-and-the-covid-19-pandemic/

Usheroff, M. (2022, March 2). *Future of COVID-19: What Comes Next?* Cedars-Sinai Medical Center. https://www.cedars-sinai.org/newsroom/future-of-covid-19-what-comes-next/#:~:text=%22By%20the%20end%20of%20the,least%20a%20few%20months%2C%20hopefully.

Weberg, D (2021). The Future of Nursing: A paradigm shifting moment. https://www.nursingworld.org/practice-policy/innovation/blog/the-future-of-nursing-a-paradigm-shifting-moment.

World Health Organization. (2022a). WHO coronavirus (COVID-19) dashboard. https://covid19.who.int/?mapFilter=vaccinations

World Health Organization. (2022b). Weekly epidemiological update on COVID-19 - 22 March 2022. World Health Organization. https://www.who.int/publications/m/item/weekly-epidemiological-update-on-covid-19---22-march

World Health Organization. (2022). WHO coronavirus (COVID-19) dashboard. https://covid19.who.int/?mapFilter=vaccinations

World Health Organization. (2022, March 22). Weekly epidemiological update on COVID-19 - 22 March 2022. World Health Organization. https://www.who.int/publications/m/item/weekly-epidemiological-update-on-covid-19---22-march-2022

World Health Organization. (2022, May). *China: WHO coronavirus disease (covid-19) dashboard with vaccination data*. WHO. https://covid19.who.int/region/wpro/country/cn

Yeung, J. (2022, March 16). *In China, 37 million people are in covid lockdown. here's what we know*. CNN. https://www.cnn.com/2022/03/15/china/china-covid-outbreak-explainer-intl-hnk/index.html

Yeung, J. (2022, April 29). *180 million people impacted by China's Covid lockdowns. here's what you need to know*. CNN. https://www.cnn.com/2022/04/28/china/china-covid-lockdown-explainer-intl-hnk/index.html

Chapter 9

Managing Digital Technology and Research Communication for Sustainable Systems

By Brendan James Moore, Syed Adeel Ahmed, and Ashraf Esmail

INTRODUCTION

Process improvements surrounding digitization and new technologies requires careful consideration by managers and leaders adopting it for their business needs, especially if the outcomes of those business drastically effect human lives, such as in the healthcare industry. In this paper, the context of surrounding the benefits digital technology has in terms of lowering cost, and aiding caregivers in communication and health outcomes is expounded, followed by field examples of implementation, and finally a discussion of challenges and naturally arising questions are discussed.

CONTEXT

In this section, the context for what digitization can do for payor and providers in the healthcare setting is discussed. Among these are that Digitization eases communication among patients and healthcare providers and patient outcomes in medical settings, digital technology can lower cost of a variety of operations, and that improvements in digitization help create sustainable environments for file management information governance. The first, however, is how digitization and analytics use can be used to ease communication of patient outcomes in medical settings. Digitization eases communication among patients and healthcare providers and patient outcomes in medical settings.

Big Data Analytics Use in Reducing Emergency Room Overcrowding

Big Data Analytics can help reduce ER (emergency room) overcrowding. By examining an ER department's case mix (what kind of cases occur at different times of day), healthcare leaders can better staff their emergency departments based on the time of day.

Overcrowded emergency rooms can happen for a variety of reasons. First, the emergency department (ED) can be understaffed. Second, there can be an unplanned large patient population size in the ER, perhaps due to large events, popular times (Friday and Saturday evenings), and natural disasters.

By ensuring the ER department has appropriate medical practitioners on their shifts, the hospital can reduce inefficiencies and create a smoother, faster service for ER patients. For example, if Mardi Gras is approaching have a contingency plan in place for an increase in admittance rates. Also, have a disaster recovery plan in place, so in the case of an emergency the hospital will be able to leverage other resources and prevent ER overcrowding and a more stable work-flow.

HOW DIGITAL TECHNOLOGY CAN LOWER COST OF A VARIETY OF OPERATIONS.

ERP Systems Reduce Cost

Concerning the success and obstacles of upgrading and implementation ERP systems, "An Enterprise Resource Planning (ERP) is business process management software that allows an organization to use a system of integrated applications to manage the business and automate many back-office functions related to technology, services and human resources" (Internal Class Document, *ERP Upgrades*, Slide 2) often discussion involves business making an eventual movement to the cloud and constantly upgrading their ERP system. However, considerations about the specific industry your business operates in is often overlooked by ERP package providers. Many ERP packages come as a single offering that does multiple functions. With this in mind, your specific industry will have hurdles and considerations that ERP implementation specialists might not consider, which make all-encompassing packages less desirable. I will use unique caveats in the healthcare industry to show that what popular articles take as general wisdom about ERP implementation and upgrades might not always apply as the best solution for a hospital system.

One challenge many hospital facilities are in rural areas, where cloud computing availability is unreliable. Because physicians and nurses often collaborate with patient sensitive data governed by HIPAA, and if a patient or doctor is unable to access patient data, even for a short duration, the health of the patient is at risk. In most industries, the consequences for being unable to access information for a short duration are not dire. Because of the health risks and legal risks of fully adopting cloud ERP capabilities, I suggest that hospitals always have a backup plan in the case of an emergency, whether that emergency is simply lack of access to the internet or a power shortage. Opting for a "hybrid" cloud and physical ERP system can allow the hospital to adopt the level of risk appropriate for being in line with HIPAA code.

Similarly, for those smaller clinics and hospitals, there is an obstacle concerning security when using machines in your business connected to the internet (the Internet of Things). Old facilitates are not used to managing a high level of data flow. Even larger community facilities, such as Saint Bernard Parish Hospital's 40 bed center use paper filing. While the paper-data storing practice of Saint Bernard Parish Hospital slows down their business intelligence practices, their security issues are narrower than a larger organization that has the resources to dedicate to a variety of data protection services.

One success of having an ERP package system is that data becomes visible across the organization. Identity management systems are expensive, and although buying an ERP package initially looks like it might save cost, your business compliance with laws and regulations, such as HIPAA, could require a level of customization that is expensive and incompatible with upgrading through the same ERP company – meaning you would have to redevelop the customization several years down the road. Before upgrading an ERP, system or implementing a new system altogether, IT managers should "point out actual problems in the system that can use improvement" and avoid buying into a system that does not consider how your business is operating in its industry (Megadatasolusi.com, p. 1).

Alternatives to buying an all-encompassing ERP package/solution that includes all features is to buy an ERP package that only includes some of the features and the rest can be done in-house. "If there are too many decision-makers involved in the planning and development of a new or modernized ERP system, the end result is likely to be a system that can do several things only moderately well – and often at a much higher cost than anticipated" (Garcia, p. 1). Thinking about your business' specific needs, and not merely differences in your business' departments, upfront will save you money and avoid buying an ERP package that fails to meet the best practices in your industry or, worse yet, buying an ERP package that is unusable by your industry's laws and regulations.

Overall, there are many success and challenges to ERP implementation as well as unique considerations for an IT manager to consider before buying an ERP package. An IT manager should consider their industry and their business' needs first and foremost before deciding on an approach to their

ERP package purchases. As demonstrated, often those packages will provide a solution many features of your business, and some are not best suited for your business' information governance practices.

Legacy Application Retirements Can Lower Costs

There are several business drivers for retiring legacy applications:

i. Cost Reduction: Even though keeping an older legacy system afloat might seem to save costs in the short term, not upgrading to a new system might cost more in the long-term, because of maintenance costs to maintain older systems.
ii. Many new applications consolidate their features into one package. For example, Adobe Suite offers yearly upgrades to all of their software in one package. If you are using a legacy application, you might have to go through several applications as a work-around to gain all of the features that a simple upgrade would solve.
iii. Many legacy applications make it difficult for a business to stay compliant with their industries regulations. Keeping up-to-date on best IT practices for your industry is important for staying in code and avoid failures during auditing.
iv. There is a risk to investing your business' success with legacy applications because your star performer who maintains the application might leave with no one in the job market who is capable of maintaining that particular legacy application. Upgrading your system and retiring a legacy application may reduce risk.

Improvements in digitization help create sustainable environments for file management information governance of Electronic Health Records. The following illustration shows how health record digitization can not only drive patient health outcomes, but it can also aid in better patient file management and a more sustainable environment than mere paper record-keeping:

In "Nurse's team writes patient stores into the EHR system," Janice Petrella Lynch demonstrates how the electronic health records system at UCLA Health System went above standard expectations in electronic health record keeping-practices to meet their patient's needs. Electronic health records display information about the patient to physicians, nurses, and other medical workers. In doing so, medical practitioners can better take care of their patient.

However, in current practice, much of patient's information history displayed is impersonal (i.e., history of blood pressure, illnesses, weight, height, etc.) and only tangentially related to the patient's narrative of their own life (the way they view their own medical history), which can leave a gap in achieving culturally sensitive, and meaningful, patient care. UCLA Health System recognized this gap in treatment and change their information gathering practices. "'The patient story report helps put the patient at the center of what we are doing,' said Ellen Pollack, MSN, RN-BC, chief nursing informatics officer, at UCLA Health System… Pollack said she loves nursing informatics [sic,] because it provides an environment that is ever changing and always challenging. 'When done right, technology can enhance our experiences and provide efficiencies," (Lynch, 2017, p.1). By adding patient's stories into EPIC, UCLA Health System was innovative with how they use currently existing technology and ensured their technology use met their patient's needs. By incorporating narrative information about the patient's history, UCLA Health System created a continuity of care, which brought impersonal information about a patient's medical history to life.

"Why did someone have an operation ten years ago?" a medical practitioner might ask. By having narrative information incorporated in UCLA Health System's use of EPIC, reasons surrounding the "why" questions become more evident. In emergency situations, I can also see how collecting narrative information can be helpful when a patient has an inability to explain their medical history.

The level of healthcare knowledge that would be required to understand this article would be an "entry-level," or basic, knowledge of healthcare, even though the article is informed by nursing informatics expert Ellen Pollack. However, having more than a basic knowledge of healthcare would enrich an understanding of the article. For example, although I may not know the specifics about how information is entered into EPIC, EHR vendors, electronic nursing documentation, bar-code medication administration, or best practices surrounding electronic health recordkeeping, I can understand the usefulness of this nursing team collecting information about patients that has been previously unsought or sought out through casual conversations with physicians and their patients, yet not recorded.

FIELD EXAMPLES

The following field examples real-world examples, such as the UCLA Patient Story case above, to illustrate benefits of digitization and digital technology use in driving process improvements to the healthcare industry. First, EHR and analytics use by vendors is discussion, followed by cloud use in record-keeping by such companies as UCHealth and RxRevu. Finally, mobile technology and recent apps, such as EQ-Active Brain Tracking, HIPPA requirements, and the Ponemon Institute study are expounded upon.

EHR USE IMPROVES ANALYTICS AND RESEARCH

Electronic Health Records (EHR) records and data analytics rely on one another. Data analytics need an EHR in order to increase data-mining capabilities, achieve statistical significance with the quantity of data gathered, and use that data to find data-driven operational improvements. Also, well-governed EHR system can save the healthcare system costs in inefficiencies, legal fees, and smoother patient transactions, access to patient records, and a greater patient experience.

Electronic Health Records (EHR) include patient health information that conforms. There are currently over 300 EHR vendors. By contrast, as we learned in class, "the extensive use of data, statistical and quantitative analysis, explanatory and predictive models, and fact-based management to drive decisions and actions" (Hoyt, 2014).

More recently, data analysts have been using Big Data to track healthcare data trends. Asking, "What kind of payor mix does your hospital encounter" and "how does your case mix affect the way you treat patients?" are important questions to meeting your customers' needs and Furthermore, "EHRs may [also] increase patient satisfaction through faster results, messaging, patient portals, electronic patient education, e-prescribing, and online scheduling (Hoyt, 2014).

DIGITIZATION OF RECORDS THROUGH THE CLOUD

A manager must consider many factors when deciding to use Cloud functions for their IT department. Thematically, they may surround data security, data handling and back-up plans, appropriate analytics, and meeting your patients/customers' needs. Some of those factors include the following:

- What kind of function is being replaced? How will that affect your patient/customers?
- How will data be backed-up?
- What is the company's plan in case of being disconnected from the Cloud functions?
- Can your organization's needs be met through another means if the service being used through the Cloud suddenly stops?
- Think about your payor mix and patients' needs. Can your patients benefit from Cloud-based subscriptions?

Using Cloud functioning for IT purposes can improve current healthcare processes. For example, UCHealth worked with a start-up vendor RxRevu to help physicians more information when prescribing drugs to patients.

> "When a provider decides to prescribe, the cloud-based decision support system, which is accessed in the electronic health records workflow, will query the EHR for patient discrete conditions and historical information in the records system that could affect the drug decision. Having the decision support tool embedded in the EHR workflow gives immediate access to the tool rather than logging in to another system" (Goedert, 2017).

However, as a manager of an IT department, you should consider all possible scenarios when choosing to use a cloud-based decision support system. For example, if the physician cannot access decision support tool due to an internet outage, will the physician still have access to enough patient conditions and historical information to make a decent diagnosis. Also, will the patient's EHR information be backed-up on-site in the event that that the hospital cannot access the information through cloud-computing. As a manager, even though using cloud computing opens many operational advantages, ensuring you have a back-up plan is essential to making sure you can meet patients' needs in the event of an outage.

The Cloud's use in healthcare is needed to meet the needs of patients in a dynamic marketplace. If patients have access to the internet and resources, moving some IT HER storage functions may allow for greater transparency between patients and their medical records. As an IT manager, think about your payor mix and your patient demographics before shifting IT functioning to a cloud-based system.

The Cloud's use in healthcare also meets the hospital's need to have health operating margins. "The global adoption for cloud services in healthcare is expected to grow from $3.73 billion in 2015 to nearly $9.5 billion by 2020" (Delaney, 2017). Besides being cheaper for organizations, cloud computing may also address opportunities for improved security if implemented correctly.

- Take the opportunity to upgrade the application to the latest release.
- Deploy the application in a tightly controlled virtual network segment.
- Introduce network-level threat prevention.
- Enforce stronger controls on underlying databases.
- Eliminate all existing server-level vulnerabilities prior to cutover" (Mellen, 2017).

While these improvements are initially tempting, security does not come automatically by moving EHR systems to the Cloud. Malware infections, and variety of hostile and intrusive software, can compromise patient data. Besides backing-up files, ensuring you know how Cloud storage works, where the data is located, and what kinds of protections are needed for

HIPAA compliance is essential to mitigating lost and mishandled sensitive patient data.

Also, be weary of what kind of standardized research data you use when using Cloud computing for analytics. For example, Oracle is now offering a data management workbench that utilizes the Cloud. "Data Management Workbench Cloud Service enables pharmaceutical companies and contract research organizations to integrate, reconcile and analyze the growing variety and volume of clinical and healthcare data" (Violino, 2017). Using these services is also appealing, since the majority of organizations say their analytics tools fall short of their needs (Violino, 2017). However, as a health organization, your IT department should know that the data being analyzed may not be indicative of your patient population. As a manager, even the IT department should think about the hospital's Case Mix (the types of cases your physicians see) when drawing conclusions through nationally aggregated, cloud-based, computed analytics. Be smart and know that conclusions drawn should be based on data from trustworthy sources. Just because a pharmaceutical company, or contracted research organization, uses Cloud computing does not mean that their results apply to your specific patient population. As an IT department manager, be sure that your own department is assisting the analytics team in making informed and sensitive decisions that cater to the organization's needs.

Overall, there are many things to consider as an IT manager when choosing to use Cloud functions for their IT department. Namely, aspects surrounding security, information handling and governance, the use of analytics, and back-up plans in case the function needs to be suddenly met without access to the cloud.

MOBILE USE FOR DIGITAL HEALTH

Mobile devices have made running applications and interacting in business environments easier than previous desktop computer-based methods. However, the convenience of using mobile devices connected to corporate networks may also create security risks that IT managers need to consider.

One consideration for mobile development is that most apps your business will use needs to have mobile access, especially if the app is customer-facing, so you must worry about whether the information will be represented clearly for mobile users. For example, "the app, called EQ-Active Brain Tracking, uses interactive mobile games to assess neurological performance to test and track neurological function." (Goedert, 2017, p.1).

Before implementing any IoT (Internet of Things) technology, you should look at a risk-benefits analysis and think about the risks associated with using IoT (Internet of Things) technology. For example, there might be an issue with hackable IoT devices, and the risks can sometimes be greater than merely financially disadvantageous outcomes: hackable Pacemakers and hackable self-driving vehicles (cars) are examples of new technology that is at risk for misuse.

The following list is of 5 popular mobile security threats that IT professionals should be aware of protecting against:

1. Data leakage
2. Social Engineering
3. Wi-Fi interference
4. Out of date devices
5. Physical Device breach

"In a 2016 Ponemon Institute study, 35 percent of professionals indicated their work devices had no mandated measures in place to secure accessible corporate data." (Raphael, p. 1). As an IT manager, you should put in place reasonable information governance principles to ensure that data leaks do not occur due to internal, as well as external, sources. Not all data leaks are malicious and taking mobile security as seriously as you take computer security is the first step in ensuring that policies governing employee mobile use are clearly understood across your company.

The healthcare industry has different privacy concerns than other industries, so mobile offerings might need extra protection that a generic package might not offer. Nurses working on the hospital "floors" might expect to access their patient's Electronic Health Records (EHR) from their phones; however, without strong information governance practices in place, what might seem like an employee innocently leaving their phone alone on a patient's bedside might actually be a violation of HIPAA (Healthcare Insurance Portability and Accountability Act).

Another example of mobile security is to be weary of other devices that interact with mobile devices that are also hackable. As discussed in class, the practice of autonomous driving vehicles might only work well if other cars are being driven by autonomous driving. However, if only a small percentage of other drivers are autonomously driving, then non-human drivers might not be able to predict how the other (human) drivers will react. Besides the obvious worries about an AI's ability to predict natural human movements and decisions, Wi-Fi interference and the vehicle's hackability also make the practice seem dangerous if implementation is not well thought out. A mobile device might be used to hack other IoT devices, and security concerns are not merely involving the mobile devices being breached.

DISCUSSION OF ISSUES

Several issues naturally arise when discussing controlling costs, mobile use, and the benefits of digitization and cloud use in communication and information governance improvements. The following discussion is to articulate what those issues and challenges are, and how to address them.

How can IT help management control exponential growth of costs for cloud computing while not being a barrier to smart use of the technology?

Although using cloud computing is cheaper than dealing with the initial start-up costs of having an in-house system, the cost is typically related to usage. Because of the direct relationship between cost and use when cloud computing, I have several suggestions that can help IT management control exponential growth of costs, while not being a barrier to cloud computing smart use.

First, if you are using cloud computing for storage, migrate as much static data as possible through a bulk load process and then continue to transmit incremental data.

Second, since resources can be dynamically reallocated per demand, make sure that your project is not duplicating efforts and that only those that need access to cloud computing at specific times are the ones using the system.

Before deciding to use cloud computing, assess the short-term as well as the long-term cost and keep in mind your usage. If you build an in-house infrastructure, you should ask yourself how much it will cost to keep up with improvements year-after-year. If you go with a cloud computing service, you should ask yourself about the costs of recovering from a disasters, such as if the service can no longer be provided. In a healthcare context, keeping the customer (or patient) in mind is fundamental in ensuring that your company's mission can be kept, regardless of which use of technology you choose to use.

WHY MOBILE TECHNOLOGY IS INCREASING IN POPULARITY AND WHAT ARE ITS CHALLENGES?

There are many reasons why mobile technology is increasing in popularity. Some of the reasons include mobile technology becoming more affordable over time. Also, mobile technology has increased in speed, convenience, wireless connectivity, and expanded in memory, while at the same time shrinking in device size.

In addition, many mobile devices are not merely used for phone calls. Instead, mobile smart phones can access the internet and use e-mail, apps, and other functions and features that increase convenience for consumers and those providing services, such as clinicians.

Many clinics use text messages to remind patients of appointments, remind patients to take medications, and to communicate lab results. With mobile technology increasing the ability for patients to have access to healthcare and other services, I do not find it surprising that mobile technology continues to increase in popularity.

Mobile technology presents unique challenges to healthcare organizations. First, if patients demand mobile technology health support, then providing that service will incur a cost. When developing an app or website that is mobile compatible, you should consider interoperability, screen size (for user ease of access) and other technical issues.

Security is also an issue. Hackers can gain access through smart devices, which include many models of mobile phones. Healthcare organizations need to consider how devices interact with one another and ensure that patient safety is a priority over convenience or patient satisfaction. For example, in the event of an operation that uses smart devices, making sure phones are off and best practices surrounding interference prevention should be implemented.

Also, if some of your treatments require sensors and mobile technology, then there are financial questions that need to be answers concerning whether or not the patient will bear the cost, or insurance companies, or the hospital system.

Mobile technology and wearables could help us move from merely treating illness to proactively reducing the occurrence of illness. Built-in analytics and dashboards enable healthcare staff and patients to sort, filter, and drill down into the patient's health data, which can be used to proactively treat and reduce the occurrence of illness.

The point is illustrated by pointing out the difference between "leading" and "lagging" indicators. For example, if a patient can know when they have an onset of a panic attack or heart problem by a notification of their vitals (or the "leading" indicators of a heart attack), they can proactively and more readily seek help, in-the-moment, rather than waiting for the "lagging" indicator of a heart attack (the phenomenological experiencing of a heart attack) in which case one would find out too late to do any proactive treatment.

Many patients are socio-economically challenged, and because of this challenge an issue is raised in healthcare. Hospitals are trying to provide the best care possible, while balancing their operational costs with the needs of the patients (cost-effective care). If a healthcare organization can send laboratory results notifications through telemedicine and make patient data available through wearables, then hospital staff and patients could save on the costs of travel, Telemedicine can also improve the bottom line of a healthcare organization, while at the same time address patients socio-economic issue. If there is a shortage of specialists in a rural area, then telemedicine may be the fastest way to provide care, while also saving costs on the patient from traveling out of their city or state to see a specialist.

How will practicing telemedicine help the healthcare organization's bottom line? If the hospital gets federal funding based on lower readmission rates, then utilizing teleconsultation and telemedicine in preventative and after-care may increase a hospital's federal reimbursement.

WHAT ARE POTENTIAL BENEFITS OF TELEMEDICINE IN DIGITAL HEALTH TO PATIENTS AND CLINICIANS?

Concerning clinicians, some benefits of telemedicine are that telemedicine allows for greater communication and improved collaboration among physicians and disparate healthcare organizations. Not all clinicians are employees of hospitals, so telemedicine specialists can be hired and contribute to healthcare organizations they otherwise would not be able to assist.

Concerning patients, teleconsultations with specialists in a variety of fields allows for greater access to care. If a patient is in a rural area with no specialists or large hospitals in their immediate areas, telemedicine allows for opportunities for care that the patients might not otherwise have. Specialties include teleradiology, teledermatology, teleneurology, telepharmacy, and many other.

Having convenient access to specialty care will likely raise patient satisfaction, which will thereby increase a hospital's HCAHPS (the Hospital Consumer Assessment of Healthcare Providers and Systems) scores, which affects federal reimbursement to the hospital system. In other words, the patient, hospital system, and clinician all can benefit from telemedicine.

Although, those using telemedicine have to keep in mind the relatively new, and changing, laws surrounding its use and also ensure that patient's needs are actually being met, regardless of patient satisfaction scores.

HOW CAN DIGITIZATION REDUCE THE "OPIOID CRISIS"?

Before elaborating on an element of HIT (Health IT) that may help reduce the "Opioid Crisis," the "Opioid Crisis" is described below:

"Over the past decade, the United States has seen a steady rise in the number of deaths caused by opioid overdoses. Although it's tough to pin down a single solution, one major factor contributing to higher overdose rates has been the over prescription of controlled pain medications. In 2010,

U.S. clinicians prescribed enough opioids to medicate every American adult around the clock for one full month" (*EPIC*, 2017).

One major elements of HIT initially become apparent that may help reduce the "Opioid Crisis." Using wearable monitoring technology to track prescription drug use may help reduce the opioid crisis. Physicians and patients alike can have greater transparency concerning drug use if data is being collected on intake.

The only drawback is that this technique might only be possible if voluntarily done since the patient may want privacy for tracking their intake. However, even if some of the patients voluntarily allow for this tracking, opioid abuse will still decrease.

WHAT ARE THE DISADVANTAGES AND ADVANTAGES OF THE CLOUD?

Disadvantages

One disadvantage of using the Cloud for storage or cloud computing is that typically you will be using a single vendor. Using a single vendor raises the question, "What if the company goes out of business?" Also, outages might occur from DoS (Denial of Service) attacks and outside threats, rather than a business closing. In case of emergencies, such as the company going out of business or a DoS attack, you should have a backup plan in place that does not require the cloud service you are considering implementing.

As a manager of an IT department, you should consider all possible scenarios when choosing to use a cloud-based decision support system. For example, if the physician cannot access decision support tool due to an internet outage, will the physician still have access to enough patient conditions and historical information to make a decent diagnosis. Also, will the patient's EHR information be backed-up on-site in the event that that the hospital cannot access the information through cloud-computing. As a manager, even though using cloud computing opens many operational advantages, ensuring you have a back-up plan is essential to making sure you can meet patients' needs in the event of an outage.

Also, HIPAA compliance is important in the medical profession. However, since cloud use is relatively new, how to meet HIPAA's standards is sometimes ambiguous, and extra effort needs to be made to ensure that you are HIPAA compliant with any cloud-based practices and personal health information (PHI) security.

In addition, you should spend extra time and effort into making sure that your cloud-based service operates across department needs and technology. If you use a different system that cannot be used by others due to technologic differences within your organizations, you may have interoperability issues. Next, we will cover some advantages of the cloud.

Advantages

Using the Cloud can also help a healthcare business in many ways. First, a growing company can more easily achieve scalability. Because cloud computing use is typically pay-per-use, a small yet fast growing company can "pay as you grow."

Because third-party cloud services are in charge of their own upkeep, typically a newer company will experience higher application performance than if they attempt to keep updated on their in-house system. Also, cloud services help have a centralized, aggregated, data source, so instead of needing to store and move data across your healthcare system, everyone can simply go to the Cloud for accessing what they need.

Concerning back-up and disaster recovery, the cloud can offer protections that paper-based EHR (Electronic Health Record) systems cannot. Patient's PHI (Personal Health Information) will not be lost in the case of a flood or natural disaster.

The Cloud's use in healthcare is needed to meet the needs of patients in a dynamic marketplace. If patients have access to the internet and resources, moving some IT HER storage functions may allow for greater transparency between patients and their medical records. As an IT manager, think about your payor mix and your patient demographics before shifting IT functioning to a cloud-based system.

The Cloud's use in healthcare also meets the hospital's need to have health operating margins. "The global adoption for cloud services in healthcare is expected to grow from $3.73 billion in 2015 to nearly $9.5 billion by 2020" (Delaney, 2017).

Also, if the healthcare company does not have the initial funds, using cloud-based services can allow companies to avoid upfront infrastructure costs and use their precious resources on other projects and business needs.

CONCLUSION

Manager and Executive Sponsorship on input and consideration surrounding digitization and digital technologies was supported. Their use and adoption requires careful consideration as to whether it benefits a business' needs, especially when the outcomes being discussed drastically effect human lives, such as in the healthcare industry. In this paper, the context of surrounding the benefits digital technology has in terms of lowering cost, and aiding caregivers in communication and health outcomes is expounded, followed by field examples of implementation, and finally a discussion of challenges and naturally arising questions are discussed.

REFERENCES

Adler-Milstein J. From Digitization to Digital Transformation: Policy Priorities for Closing the Gap. JAMA. 2021;325(8):717–718. doi:10.1001/jama.2020.27014

Adu, Ernest Kwadwo; Mills, Annette; and Todorova, Nelly, "The Digitization of Healthcare: Understanding Personal Health Information Disclosure by Consumers in Developing Countries - An Extended Privacy Calculus Perspective" (2017). ACIS 2017 Proceedings. 111. https://aisel.aisnet.org/acis2017/111

Ahlin, K.; Snyder, K. (2021). Person-centred Health Care Resting on Digitization and Systematic Processes : A position paper. *GLOBAL HEALTH 2021 The Tenth International Conference on Global Health Challenges*. Presented at the GLOBAL HEALTH 2021 The Tenth International Conference on Global Health Challenges, Barcelona, October 3-7, 2021. Retrieved from http://urn.kb.se/resolve?urn=urn:nbn:se:miun:diva-43461

Aminololama-Shakeri S, López JE. The doctor-patient relationship with artificial intelligence. *AJR Am J Roentgenol*. 2019;212(2):308-310. doi:10.2214/ajr.18.20509

Arora A. Conceptualising artificial intelligence as a digital healthcare innovation: an introductory review. *Med Devices (Auckl)*. 2020;13:223-230. doi:10.2147/mder.s262590

Balthazar P, Harri P, Prater A, Safdar NM. Protecting your patients' interests in the era of big data, artificial intelligence, and predictive analytics. *J Am Coll Radiol*. 2018;15(3 Pt B):580-586. doi:10.1016/j.jacr.2017.11.035

Birdas TJ, Rozycki GF, Dunnington GL, Stevens L, Liali V, Schmidt CM. "Show Me the Data": a recipe for quality improvement success in an academic surgical department. *J Am Coll Surg*. 2019;228(4):368-373. doi:10.1016/j.jamcollsurg.2018.12.013

Bloomberg News, 2017. "Microsoft tries again in healthcare, this time with cloud, AI" Health Data Management, February 16th.

Campbell EG. Doctors and drug companies-scrutinizing influential relationships. *N Engl J Med*. 2007;357(18):1796-1797. doi:10.1056/NEJMp078141

Carter SM, Rogers W, Win KT, Frazer H, Richards B, Houssami N. The ethical, legal, and social implications of using artificial intelligence systems in breast cancer care. *Breast*. 2020;49:25-32. doi:10.1016/j.breast.2019.10.001

Coglianese C. The transparency president? the Obama administration and open government. *Governance*. 2009;22(4):529-544. doi:10.1111/j.1468-0491.2009.01451.x

Delaney, D., 2017. "Why healthcare is continuing its shift to the cloud" Health Data Management, April 27th.

Dowie J, Kaltoft MK. The future of health is self-production and co-creation based on apomediative decision support. *Med Sci (Basel)*. 2018;6(3):66. doi:10.3390/medsci6030066

EPIC Systems Corporation, 2017. "Curbing Opioid Abuse: A Three-Pronged Approach," Success at Seven, Metro Health, v.5; issue 1.

ERP News, 2018. "ERP Strategy: When is it Time to Upgrade?"

Frow P, McColl-Kennedy JR, Payne A. Co-creation practices: their role in shaping a health care ecosystem. *Ind Mark Manag*. 2016;56:24-39. doi:10.1016/j.indmarman.2016.03.007

Fukami T, Uemura M, Nagao Y. Significance of incident reports by medical doctors for organizational transparency and driving forces for patient safety. *Patient Saf Surg*. 2020;14:13. doi:10.1186/s13037-020-00240-y

Garcia, T., 2014. "3 Rules for ERP Implementation: Reflect on these considerations when applying or upgrading an enterprise resource planning system" *Material Handling and Logistics*, Technology and Automation section.

Goedert, J., 2017. "Decision support gives UCHealth physicians more treatment options" Health Data Management, June 8th.

Goedert, J. 2018. "Mobile app tracks neurological performance," HealthDataManagement, November 1.

Goupil B, Balusson F, Naudet F, et al. Association between gifts from pharmaceutical companies to French general practitioners and their drug prescribing patterns in 2016: retrospective study using the French Transparency in Healthcare and National Health Data System databases. *BMJ*. 2019;367:l6015. doi:10.1136/bmj.l6015

Gunning D, Stefik M, Choi J, Miller T, Stumpf S, Yang GZ. XAI-Explainable artificial intelligence. *Sci Robot*. 2019;4(37):eaay7120. doi:10.1126/scirobotics.aay7120

Habli I, Lawton T, Porter Z. Artificial intelligence in health care: accountability and safety. *Bull World Health Organ*. 2020;98(4):251-256. doi:10.2471/blt.19.237487

Hajjo R. The Ethical Challenges of Applying Machine Learning and Artificial Intelligence in Cancer Care. In: 2018 1st International Conference on Cancer Care Informatics (CCI); November 19-21, 2018; Amman, Jordan. doi:10.1109/cancercare.2018.8618186

Hoyt, R.E., 2014. *Health Informatics: Practical Guide for Healthcare and Information Technology Professionals 6th edition*, Informatics Education, June.

Hughes G. *The Promise of Conversational AI in Helping Restore the Doctor-Patient Relationship*. MedCity News; 2020.

Jiang F, Jiang Y, Zhi H, et al. Artificial intelligence in healthcare: past, present and future. *Stroke Vasc Neurol*. 2017;2(4):230-243. doi:10.1136/svn-2017-000101

JMIR Med Inform 2022;10(1):e28981 URL: https://medinform.jmir.org/2022/1/e28981 DOI: 10.2196/28981

Khan ZH, Siddique A, Lee CW. Robotics Utilization for Healthcare Digitization in Global COVID-19 Management. International Journal of Environmental Research and Public Health. 2020; 17(11):3819. https://doi.org/10.3390/ijerph17113819

Kostkova P, Brewer H, de Lusignan S, et al. Who owns the data? open data for healthcare. *Front Public Health*. 2016;4:7. doi:10.3389/fpubh.2016.00007

Kuan R. Adopting AI in health care will be slow and difficult. *Harv Bus Rev*. 2019 October 18, 2019. https://hbr.org/2019/10/adopting-ai-in-health-care-will-be-slow-and-difficult. Accessed January 26, 2020.

Lalmuanawma S, Hussain J, Chhakchhuak L. Applications of machine learning and artificial intelligence for Covid-19 (SARS-CoV-2) pandemic: a review. *Chaos Solitons Fractals*. 2020;139:110059. doi:10.1016/j.chaos.2020.110059

Lauritsen SM, Kristensen M, Olsen MV, et al. Explainable artificial intelligence model to predict acute critical illness from electronic health records. *Nat Commun*. 2020;11(1):3852. doi:10.1038/s41467-020-17431-x

Lee D. Effects of key value co-creation elements in the healthcare system: focusing on technology applications. *Serv Bus*. 2019;13(2):389-417. doi:10.1007/s11628-018-00388-9

Lehne M, Sass J, Essenwanger A, Schepers J, Thun S. Why digital medicine depends on interoperability. *NPJ Digit Med*. 2019;2:79. doi:10.1038/s41746-019-0158-1

Leone D, Schiavone F, Appio FP, Chiao B. How does artificial intelligence enable and enhance value co-creation in industrial markets? an exploratory case study in the healthcare ecosystem. *J Bus Res*. 2020. doi:10.1016/j.jbusres.2020.11.008

Levay C, Waks C. Professions, and the pursuit of transparency in healthcare: two cases of soft autonomy. *Organ Stud*. 2009;30(5):509-527. doi:10.1177/0170840609104396

Lusch RF, Vargo SL, Tanniru M. Service, value networks and learning. *J Acad Mark Sci*. 2010;38(1):19-31. doi:10.1007/s11747-008-0131-z

Ly S, Tsang R, Ho K. Patient Perspectives on the Digitization of Personal Health Information in the Emergency Department: Mixed Methods Study During the COVID-19 Pandemic

Lynch, J. P., MSN, RN, "Nurse's team writes patient stores into the EHR system," nurse.com, January 19th, 2017.

Megadatasolusi.com, 2018. "Best Way to Present ERP System Upgrade Case to Higher Management"

Mellen, M, 2017. "4 reasons to prioritize cloud security this year" Health Data Management, February 16th.

Nagy M, Sisk B. How will artificial intelligence affect patient-clinician relationships? *AMA J Ethics*. 2020;22(5):E395-400. doi:10.1001/amajethics.2020.395

Nundy S, Montgomery T, Wachter RM. Promoting trust between patients and physicians in the era of artificial intelligence. *JAMA*. 2019. doi:10.1001/jama.2018.20563

Oettgen P. Transparency in Healthcare. EBSCO Health, DynaMed Plus. https://www.ebsco.com/sites/g/files/nabnos191/files/acquiadam-assets/66751087.pdf.

Pronovost PJ, Wu AW, Austin JM. Time for transparent standards in quality reporting by health care organizations. *JAMA*. 2017;318(8):701-702. doi:10.1001/jama.2017.10124

Selbst AD, Powles J. Meaningful information, and the right to explanation. *Int Data Priv Law*. 2017;7(4):233-242. doi:10.1093/idpl/ipx022

Spatharou A, Hieronimus S, Jenkins J. *Transforming Healthcare with AI: The Impact on the Workforce and Organizations*. https://www.mckinsey.com/industries/healthcare-systems-and-services/our-insights/transforming-healthcare-with-ai. Published 2020.

Steinhubl S, Topol E, et al. Moving From Digitalization to Digitization in Cardiovascular Care. J Am Coll Cardiol. 2015 Sep, 66 (13) 1489–1496. https://doi.org/10.1016/j.jacc.2015.08.006

Violino B., 2017. "Spending on cloud computing soars during 2016" Health Data Management, January 17th.

Violino B., 2017. "Analytics initiatives are encouraging cloud adoption" Health Data Management, May 25th.

Violino, B., 2017. "Majority of organizations say analytics tools fall short" Health Data Management, June 6th.

Violino B., 2017. "Oracle releases Data Management Workbench Cloud Services" Health Data Management, April 27th

Wachter RM, Howell MD. Resolving the Productivity Paradox of Health Information Technology: A Time for Optimism. JAMA. 2018;320(1):25–26. doi:10.1001/jama.2018.5605

Yasanur Kayikci, Sustainability impact of digitization in logistics, Procedia Manufacturing, Volume 21, 2018, Pages 782-789, ISSN 2351-9789, https://doi.org/10.1016/j.promfg.2018.02.184.

Chapter 10

Harnessing Global Energy Mix

The African Perspective

By Edward Agbai and Theophilus Boufini

INTRODUCTION

The current population of Africa is estimated to be about 1.3 billion people in 2021, based on the latest United Nations estimates. Africa's population is equivalent to 16.72% of the total world population of about 7.9 billion people. The continent is ranked number two among regions of the world by population, while Europe, with about 0.75 billion people, is equivalent to 9.78% of the world population. While Africa's contribution to global greenhouse gas emissions is about 10 percent, Europe contributes about 8 percent to Global Greenhouse Gas Emissions (GHGE). If we must make a simplified comparison, a person living in Africa contributes less than a person in Europe. We need to establish this context for proper analysis of the effects of energy mix discourse and the need to have a proactive leadership and governance system globally.

Despite the average African contributing less to the GHGE, we all share a common humanity that needs sustainability to exist. The adverse impact of global warming is evident for all to see. In 2020, Africa was devastated by Cyclone Idai, which ravaged countries in southern Africa, like Malawi, Mozambique, and Zimbabwe. Families were made homeless, and hundreds of thousands of children needed humanitarian assistance. While these countries were battling to recover their lives from the destruction, cyclone Kenneth hit Mozambique and brought powerful winds and heavy rains, leading to further devastation. According to a report by UNICEF, these two cyclones displaced thousand-folds of families, leaving many in need of basic supplies like food and water and bringing widespread flooding and destruction of almost 780,000 hectares of crops (UNICEF, 2020)

There is no doubt that because of its high exposure and vulnerability to climate hazards, a third of the people considered most at risk from climate change live in Africa. A recent report by McKinsey Sustainability in October 2021 titled "Green Africa: A growth and resilience agenda for the continent" indicated that "More than 900 million inhabitants of the African continent could find themselves exposed to one or more climate hazards in a warming world over the next 30 years. Simultaneously, shifting global consumption patterns as the world moves towards a low-carbon future—including a global fall in demand for oil could further undermine African economies." These realities have made African leaders and governments marshal out policies to mitigate these risks. It, therefore, became imperative for African leaders and experts in charge of governance to attend the 26th session of the Conference of the Parties (COP 26) to the UNFCCC, held in November 2021, in Glasgow, UK. The plan centered on finance, gender, youth, and public empowerment. Many speakers focused on these dominant themes at the conference, which saw a collective human endeavor to stabilize the climate.

The forum provided the African governments with the opportunity to understand the overarching agenda of net-zero world policies of COP26. The conference provided the platform for the continent leaders to use its vast agricultural landscape for a "Green Africa" to use alternative forms of energy available in renewable energy and expand its renewable energy generation capacity. The continent has been known for its natural resource endowments, but the leaders need to change the narrative to a rich natural human capital endowment. This assertion is to guarantee responsible and strategic leadership and governance that is sustainable to catalyze economic growth and make a substantial contribution to the global net-zero transition.

A decarbonized environment is a sustainable environment. The COP 26 put global efforts at decarbonization in focus. It is worth mentioning that developed markets have made progress in meeting the challenge of climate change; the sequencing and the transition to a greener future are critical. African countries must tap into it and access international support to pursue low-carbon development pathways and build a safer, healthier, and more prosperous future for their citizens.

A decarbonized environment puts pressure on access to affordable energy, becoming a global concern. Priority to promote and facilitate access to renewable energy sources needs to be expedited. Nigeria, the most populous African country, has intensified efforts to migrate from the premium motor spirit and diesel to natural gases as the main source of energy and increase the availability of clean energy. Nigeria possesses significant renewable energy potential, including solar, wind, and tidal energy. The country depends highly on fossil fuels to supply energy demand, especially since financing renewable energy projects brings some specific challenges that need to be addressed.

This chapter will explore the background of the energy crisis, what is causing the recent global energy crisis, and possible solutions to the crisis. It will also explore the energy transition by highlighting the sources of renewable energy, legislations, policy considerations, what leaders are doing, and emerging trends in the energy mix in Africa. The chapter will close with what it means to own an electric car in Africa.

THE CONTEXT OF ENERGY MIX

There is the need to encourage the development of innovative homegrown goals for country-specific and organization-centered energy transition both at the micro-level and global energy transition at the macro level for development partners.

We live in times of rising energy demand occasioned by the crisis between Russia and Ukraine, but the crisis did not start today. It began when the Arab producers of the Organization of Petroleum Exporting Countries (OPEC) put an embargo on oil exports to the United States in October 1973 and threatened to cutback overall production by 25 percent. This resulted in a price increase in the international oil and gas market. The crisis lasted for six months before the embargo was lifted in March 1974 after negotiations at the Washington Oil Summit. However, the effect of the crisis lingered throughout the 1970s. In 1977, the Nigerian government declared a festival of arts and culture tagged "Festac 77' from the proceeds of the oil windfall.

The crisis of the 1970s prompted the first shift towards energy-saving technologies. It made policymakers globally think deeply about the future of energy within the newfound power of the oil-producing countries, who used the crisis to find their voices on the world stage. It brought new dynamics to the geopolitical equation of the world's major industrial centers, as developed economies had to contend with escalating issues related to petroleum supply. The new reality because of the crisis led to a stagnant economic outlook worldwide and affected the developed economies. Fossil fuels have driven the world economy and supported the second and third industrial revolutions for over a century. The sources of fossil fuel are not renewable and unsustainable in the long run, with reserves depleting very fast. The three primary fossil fuels: oil, coal, and natural gas, provided fuel for over 80% of world energy consumption in 2019. Fossil fuel will be relied upon to drive economic growth in the short run as the energy transition is not catching up fast as desired.

WHAT IS CAUSING THE RECENT GLOBAL ENERGY CRISIS

We were greeted in Q4 2021 with the threat by Russia to invade Ukraine, and it was the dominant news for the most significant part of Q1 2022. The proposed invasion triggered tension across the region, and Russia using the pretext of the invasion, weaponized gas supply to most of Western Europe. Russia actualized her threat and invaded Ukraine on February 24, 2022. This led to embargos and economic sanctions on Russia by the United States and the European Union. Consequently, oil and gas supply were affected. The gas supply shortage was partly responsible for the massive spike in gas prices. The crisis has had wide-ranging impacts on economics, the environment, and security. The gas supply-induced crisis had three distinct features: COVID-19 and supply chain disruptions, greater interdependence of natural gas markets, and signs of energy price volatility during the energy transition away from premium motor spirit and diesel among developed economies.

Let us not forget that the pandemic brought about a historic drop in energy demand and prices in February 2020 as movement across the world was halted for over six months. The drop in demand and prices continued up to Q2 2021; however, in Q3 2021, countries started opening their borders, which led to an increase in the demand for fossil fuels. Fast-forward to Q1 2022, recovering demand is straining market demands. Prices are burgeoning as demand chases fuel supply that has not yet recovered from the pandemic drop.

Pre-pandemic supply crises have typically been limited to unhindered oil with gas and coal supplies. Nevertheless, this is not the case in the pandemic era, as natural gas markets are also in crisis. The change in demand quotient could be attributable to a growing and more flexible liquified natural gas (LNG) market that has enabled global competition for gas supply. We can situate the reasons around the mode of transportation of the product. The gas was initially supplied by pipelines or LNG under long-term contracts. North American and European countries mainly used natural gas; however, today, Asia competes for the same LNG supply. The natural law of demand and supply comes to play, effectively pushing prices up in both markets to benefit the producing nations.

The demand for increased use of natural gas was in response to increased demand for the energy transition to cleaner forms and ultimately migrate to renewable energy. Natural gas was preferred ahead of coal in the energy transition because its burns more efficiently with less carbon released into the atmosphere. Natural gas has become a casualty of its success and effectively displacing coal-fired power generation for environmental reasons.

The global energy markets are naturally price-inelastic and thus volatile. Nonetheless, the recent emphasis on the environment and affordability during the early energy transition may have led to less attention to energy security (Gilbert et al., 2021; Zappa et al., 2021). The new interdependence of energy markets across sources and locations has also changed how the crises spread. Measures like strategic reserves and demand response may need more attention and programs to help lower-income consumers, who are always most affected when high energy prices. Diversifying the energy supply with renewables will also help, as once built, these sources are not subject to the whims of global markets (Demırbas, 2017; Newbery, 2016).

POSSIBLE SOLUTIONS TO THE PROBLEM OF GLOBAL ENERGY CRISIS

Many possible solutions are already in place today, but they have not been widely adopted:

1. Move Towards Renewable Resources - The best possible solution is to reduce dependence on non-renewable sources of energy and improve overall conservation efforts. Much of the industrial age was created using fossil fuels, but proven technology using renewable sources like steam, solar, and wind has gained acceptance.
2. Buy Energy-Efficient Products - Replace traditional bulbs with CFLs and LEDs. They use less wattage of electricity and last longer. Increased usage of these products by a more significant percentage of the world population for residential and commercial purposes will reduce the energy demand.
3. Lighting Controls - Several new technologies make lighting controls much more exciting and save lots of energy. Pre-set lighting controls, slide lighting, touch dimmers, and integrated lighting controls are a few lighting controls that can help conserve energy and reduce overall lighting costs.
4. Easier Grid Access - People who use different options to generate power must be permitted to plug into the grid and get credit for the power feed. The hassles of getting credit for supplying surplus power back into the grid should be removed. Besides, subsidies on solar panels should be given to encourage more people to explore renewable options.
5. Energy Simulation - Big corporates and corporations can use energy simulation software to redesign the building unit and reduce energy costs. New buildings with power-saving features and heat-conserving materials are used in North America, especially Canada, to conserve heat during

the cold winter months that last up to six months in some provinces. Such efforts lead to energy-efficient buildings and reduce carbon footprint.

6. Perform Energy Audit - Performing an energy audit will help identify energy wastage /loss areas. The audit will go a step further to recommend what can be done to improve energy efficiency. When done by a professional, an energy audit can reduce carbon footprint, save energy and money, and avoid energy crises. Most developed countries place a high premium on carbon credit to encourage citizens to reduce greenhouse gas emissions.
7. Common Stand on Climate Change-Both developed and developing countries should adopt a joint stand on climate change. They should reduce greenhouse gas emissions through a practical cross-border mechanism. The population growth pattern is not sustainable due to overconsumption of resources, with dire consequences of global warming and climate change (Buhaug et al., 2008; Ganivet, 2020). Countries across the board (developed and developing countries must focus on emissions cuts to halve current emissions levels by 2050 (Nadel & Ungar, 2019; Skea & Nishioka, 2008).

SOURCES OF RENEWABLE ENERGY

Wind Energy

Wind power is a renewable form of energy that describes how the wind is used to generate electricity. Power source from electricity generated from wind energy is one of the fastest-growing electrical generation methods in the world. Wind turbines mounted in locations where there are favorable weather patterns convert the kinetic energy from moving air into electricity.

It provides further alternatives for harnessing wind power. 93 GW of wind power will be installed in 2020, with US and China leading, according to the global wind report. It is the greenest power source with the most decarbonization potential/megawatt. However, it is for large-scale energy needs. Wind power has the advantage of cost-effectiveness, creates jobs, is sustainable, and can be built on existing farms or ranches. It has the drawback of unpredictability, threatens wildlife, and creates low-level noise, with limited locations suitable for wind turbines.

Solar Power

Solar energy is the radiation from the sun capable of producing heat, causing chemical reactions, or generating electricity. The total amount of solar energy received on Earth is vastly more than the world's current and anticipated energy requirements. Solar panels harness the energy from the sun and convert it into electric power. They are mounted on rooftops, on open fields, or in open spaces around the house. Solar energy has been touted as one way to effectively reduce our demand for fossil fuels, limit greenhouse gas emissions, and shrink our carbon footprint. We can broadly categorize the type of power according to how closely it is connected to traditional power utility infrastructure (grid). Within these categorizations, we can distinguish three residential solar electric power systems: the grid inter-tied, grid inter-tied with battery backup, and off-grid.

The main disadvantages of solar energy are dependence on weather conditions and the inability to store electricity. Solar energy output depends mostly on direct sunlight. A cloudy day can reduce electricity generation by more than 80%. However, a combination of solar and batteries becomes an excellent choice. It provides clean and renewable energy as the homes are emissions-free. Solar energy reduces the user's carbon footprint, saves money on electricity bills, and increases home value.

Wind and solar power as renewables offer different challenges; they complicate grid management because they are variable. The source of their variability stems from the unreliability of their sources of energy. They come and go with the weather; experts cannot ramp them up and down like fossil fuel plants. The sun comes up; solar panels flood power to electrical appliances and gadgets; the sun goes down, and none is available. These unreliable sources put stress on the electric grid. While these solar panels and wind turbines have roles in the global energy mix, they cannot be relied on for steady power streams.

According to World Energy and Climate Statistics Yearbook (2021), in 2020, the share of wind and solar in the global power mix rose by 1.2 pts, as renewable power generation continued to grow at a high pace (+12% for wind and +20% for solar), while thermal – especially coal-fired – and nuclear power generation declined.

A great percentage of this increase occurred in four countries (Belgium, China, Germany, and the United Kingdom) that invested in offshore wind power generation and accounted for an increase of 16%. The countries added a combined 20GW over five years leading up to 2020. Around the same period, the share of renewable energy sources within the global power generation mix rose by 1.8 percentage points to over 28% of the power mix. Favorable hydro conditions raised renewable power generation in China, Russia, Europe, Brazil, and Japan.

Fossil fuels are expected to meet much of the world's energy demand despite renewable energy and nuclear power growth. To meet increasing demand, countries will need to explore and adopt advanced technology to maximize production yields and increase oil and gas proven reserves to monetize and gain optimal value from their resource assets. The energy transition regime poses a challenge and opportunity for resource-rich nations to reposition their energy priorities. Such priorities are given direction through regulatory policies to develop clean and renewable energy. Industry investors are tweaking their needs assessment to provide a green energy roadmap and green story as a basis for attracting the required funds for development. Energy transition should serve as an enabler to create access to funding for fossil resources.

LEGISLATING THE ENERGY MIX

The attempt by the elite to hold on to the current is coming at a high cost to the collapse of critical social and environmental systems and threatening the survival of the human species and our civilization. The realities of the now avail us the rare opportunity to end this era of dominance of the fossil fuel power industrial revolution of the last century. The realities of deliberately legislating the energy mix are attributed to the twin action of the depletion of fossil fuel reserves and its unsustainable future of powering the global economy. The need is also anchored on expanding the transportation and communication sectors that are the main consumers of energy. Thereby reminding everyone not to walk away from the false premise that things will remain the same. The opportunity presented by COP26 is that one-way leaders can engage in turning our culture, economy, and politics from dominance to a partnership relationship. The UN Climate Change Conference, COP26, provided an essential platform for businesses and organizations to demonstrate how they act and mobilize finance to support the transition to net-zero.

COP has a historical basis; since 1995, the countries bound by the UN Framework Convention on Climate Change (UNFCCC) have been at the forefront of advocating for a change in attitude by both political and business leaders. Various COPs preceding COP26 had produced action plans. These

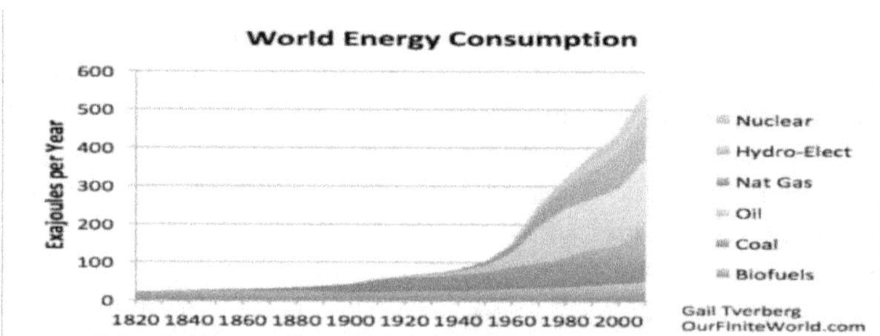

Figure 10.1 World Energy Consumption. https://oilprice.com/Energy/Energy-General/The-Energy-Crisis-That-No-One-Is-Talking-About.html

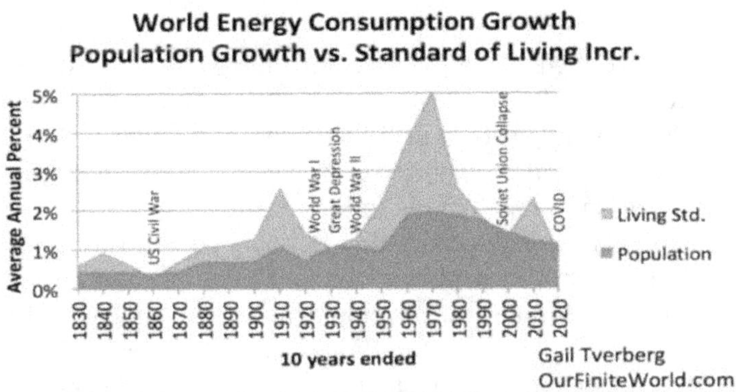

Figure 10.2 World Energy Consumption Growth-Population Growth Vs Standard of Living Incr. https://oilprice.com/Energy/Energy-General/The-Energy-Crisis-That-No-One-Is-Talking-About.html

action plans like Bali, the 2007 actions plan, the 2011 Dubai platform, the 1997 Kyoto protocols, and the 1995 Berlin mandates. Discussions in the conferences have been intense and frank. Still, we have also had acrimonious breakdowns of discussions like the Copenhagen, 2009 and the Paris agreements of 2015, which the former President of the USA, Donald Trump, vehemently opposed and tried to scuttle the efforts of that accord. But the rise in the atmosphere's greenhouse gas content and the associated warming of the climate continues despite them, even when they are often hyped as the world's last chance.

The need to legislate and put action to words is important, people of the Niger Delta, region of Nigeria been victims of the devastating effects of global warming during the 2012 flooding of the plains of the Delta. Houses in Delta state, south-south Nigeria, were covered with floodwater up to the lintel level for over six weeks; people Kpakiama community in Bomadi Local Government Area had to take refuge in neighboring communities on a higher plane. The access road to my community was washed away by the flood, and the properties destroyed were worth over 200,000.00 USD. Such flood levels have not been witnessed in my lifetime, and records show that it has not occurred in such magnitude for over 60 years.

RENEWABLE ENERGY: CATALYZING A SUSTAINABLE FUTURE FOR NIGERIA

Transitioning to Renewable Energy

Several developing countries have been striving to implement renewable energy initiatives and solutions for over a decade. Such efforts can be seen in

Nigeria, a resource-rich but energy-poor nation. Nigeria is transitioning from PMS and AGO to natural gas as a component of transitioning to renewable energy and is bolstering this with the accompanying legislative developments to optimize renewable energy and energy efficiency fully. Nigeria has an ambitious plan of working towards 100% and, in the future, 300% renewables. Surplus energy will be exported after meeting domestic needs. Nigeria aims to become a renewable training hub for Africa.

Nigeria has shown an everyday reality in many countries: transitioning to renewable energy cannot be done without active collaboration from partners like the United Nations Industrial Development Organization (UNIDO). Multinational organizations have spent millions of dollars in research and project development to understand the inherent potential of clean energy. However, an unfortunate reality in Nigeria is that the barrier to effective transitioning and financing of potential projects often lacks clear objectives and implementation strategies.

POLICY CONSIDERATIONS FOR RENEWABLE ENERGY

Energy poverty

Energy poverty lacks access to modern energy services like household access to electricity and clean cooking facilities (IEA). Table 10.1 shows the aspects that must be considered when implementing effective policies and, therefore, creating more financially viable renewable energy projects in Nigeria and other developing countries. These issues need to be addressed to secure renewable financing, create multi-stakeholder involvement, and create consistent data. These cross-sectoral policies require essential collaboration to impact the market positively. To have a smooth energy transition; developing countries need to do the following:

- Promote timely collection of reliable, and detailed data on energy end uses and market technologies.
- Iteratively update strategies and policies based on up-to-date evidence.
- Minimize market distortions caused by subsidies and other disincentives to energy efficiency and renewable energy.
- Facilitate private investments in energy efficiency

Additionally, strengthening homegrown capacity, strengthening coordination and planning mechanisms between stakeholders, and identifying the most suitable financing mechanisms for energy efficiency.

Financing Renewable Energy Projects in Nigeria

The ministries of Power, Water Resources, and Finance are government ministries saddled with the responsibility of superintending electricity and water resources in Nigeria. These ministries provide insights into relevant aspects of funding large and small-scale renewable projects. Some of the insights include:

The Challenges of Renewable Energy Projects

The significant challenges of renewable energy include issues with regulatory frameworks and policies. Regulatory frameworks can be problematic because many electrical utilities in Nigeria monopolize the electricity value chain from generation to transmission. Furthermore, it has not implemented 100% metering of electricity customers to encourage private sector participation. Lack of transparency and its accompanying lack of expertise are significant obstacles that lead to unclear implementation roadmap and unsound monitoring and reporting. These practical challenges in project development limit access to consistent funding and create gaps in data. Nigeria has organized loans, tax exemptions, and other measures to incentivize the widespread use of residential renewable energy technologies to mitigate these challenges.

An Example of a Renewable Energy Project

A micro solar photovoltaic power plant was constructed by the Niger Delta Development Commission (NDDC) in Kpakaima town, Bomadi local government area, Delta State, Nigeria, currently supplies 39,3 KWp for 24 hours a day to 75 houses and businesses. Fishermen use the excess energy produced in the photovoltaic power plant in the rural community for refrigeration purposes. This case study illustrates that renewable energy is indeed a worthwhile endeavor that facilitates the implementation of SDG #7, affordable and clean energy. Such a project is profoundly beneficial and will create a more sustainable future.

ADOPTING NATURAL GAS AS TRANSITION FUEL: THE NIGERIA EXAMPLE

The global energy transition and the need to remain relevant in the energy value chain make countries think differently. The energy sector of every nation is pivotal to rapid industrialization and development for a sustainable future. The energy sector should achieve the gold standard status with many giant leaps. Expectations and optimism are high among stakeholders of the energy sector in Africa. Industry players are supercharged to meet the high expectations of the energy demand within the broader context of a growing and active population. In the last decade, a population has witnessed violent crimes, with pockets of civil unrest across the African landscape.

Nigeria is a major natural gas producer and is caught up in the energy transition to a cleaner energy dilemma. However, it is not oblivious of its vast natural gas reserves and what it means to its growth. The Nigerian government's Decade of Gas initiative (2021 -2030) is an initiative through which the Federal government has declared natural gas as a transition fuel for green energy. The decade of gas links natural gas with more sustainable energy sources. The country is aware of its vast hydrocarbon resources but uses the Decade of Gas initiative to remind itself that the global energy transition is accurate and on course. The government is aware that the energy transition is the direction the global energy sector is moving, and the country cannot act differently. The country passed its Petroleum Industry Act (2021) to reemphasize the point of maximum petroleum product utilization within the energy transition window.

Climate change has brought discourse on energy transition to the front burner. The energy transition is to reduce carbon footprint to ensure temperatures are within allowable limits of 1.5 degrees centigrade. The African continent is bedeviled with energy poverty, a critical enabler for achieving the 17 United Nations sustainable development goals (SDGs). The goals set out in the 2020-2030 agenda for sustainable development, the 2015 Paris agreement on climate change, and Glasgow COP 26 provide an integrated pathway towards a more prosperous, equitable, and sustainable future for all. A future enabled by a just and inclusive transition to a clean and modern energy mix. Both energy poverty and climate change nexus must be addressed simultaneously in adopting future global fuels, which a robust energy mix provides. The future of global fuels must be geared toward clean and sustainable sources for resources-rich countries such as Nigeria. Nigeria represents in 2022 a paradox, a resource-rich but energy-poor nation, the country in March 2022 witnessed the high cost of refined petroleum products at the pumps with all four products of Dual-Purpose Kerosene (DPK), Automotive Gas Oil (AGO), Premium Motor Spirit (PMS) and Jet Fuel. The price doubles within two months from a January 2022 average of 288 Naira ($.70) per liter of diesel to over N700 ($1.69). The diesel example reflected the price increase of other petroleum products fueling the Nigerian economy and caused widespread anxiety for businesses that rely on the products to power generators that make up for the historically unstable power supply (Onukwe, 2022). The situation was not different in Canada, as the price rose over 50 percent within the period under review.

Table 10.1 Energy Challenges in Nigeria

Technical	*Socioeconomic*
Low power generation capacity	Unstable and unreliable tariff regime
Inability to meet existing and future energy needs	Vulnerability to the global volatile price of fossil fuel
Obsolete equipment	Energy poverty

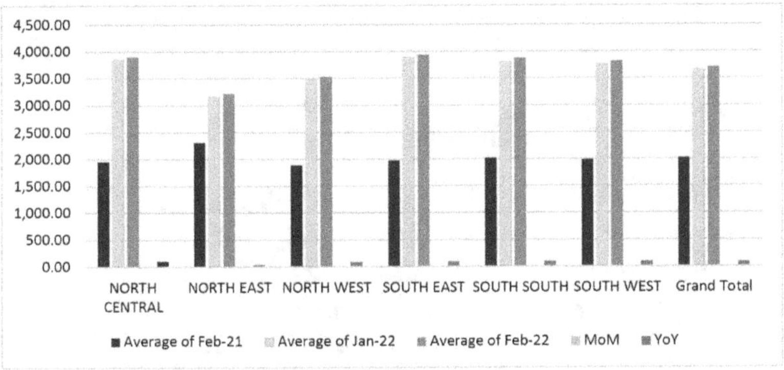

Figure 10.3 Liquefied Petroleum Gas (Cooking Gas) Price Watch in Nigeria February 2022 https://www.nigerianstat.gov.ng/pdfuploads/LPG%20FEBRUARY%202022%20REPORT.pdf

In both resource-rich and energy-poor countries, the energy transition must not come at the expense of affordable and reliable energy for citizens, cities, and industries. Instead, the energy transition must be equitable, inclusive, and just. Such transition should necessitate preserving sustainable development and poverty eradication as enshrined in global treaties such as the Glasgow COP 26. What makes the most sense from the point of view of balancing energy security with environmental sustainability is gas as a transition fuel for resources-rich countries but different for non-resource countries.

For many African countries with substantial infrastructural deficits, it makes sense to drive infrastructural and industrial development to prosper the generality of their citizenry and make life more meaningful. Natural gas, which is abundant in some African countries like Mozambique and Angola, ticks all the boxes as a vehicle to help the government of African countries achieves their aspiration. The countries are encouraged to embrace natural gas as a resource to turn around the fortune of their people. For these countries to demonstrate commitment to an efficient energy transition using gas, they can embark on a critical pathway to ensuring that the abundant gas is marshaled to engender economic growth and development. In driving to make natural gas her dominant energy source, they can follow the Nigerian

Table 10.2 Liquefied Petroleum Gas (Cooking Gas) Price Watch (February 2022)

	Ave of Feb-21	*Aver of Jan-22*	*Ave of Feb-22*	*MoM*	*YoY*
North Central	1,955.07	3,856.87	3,907.07	1.30	99.84
Northeast	2,309.30	3,166.33	3,219.75	1.69	39.43
Northwest	1,890.41	3,494.37	3,532.47	1.09	86.86
Southeast	1,977.25	3,896.03	3,938.00	1.08	99.17
South-South	2,022.69	3,805.47	3,875.23	1.83	91.59
Southwest	1,983.82	3,760.12	3,813.49	1.42	92.23
Grand Total	2,018.91	3,657.57	3,708.58	1.39	83.69

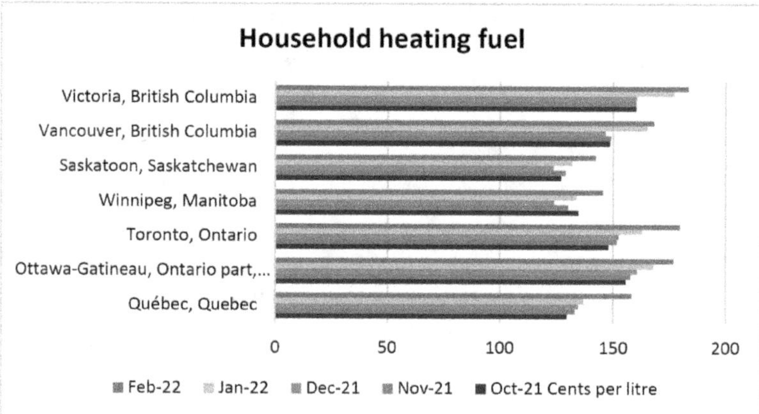

Figure 10.4 Household Heating Fuel in Canada October 2021-February 2022 https://www150.statcan.gc.ca/t1/tbl1/en/tv.action?pid=1810000101

example of making natural gas her primary energy source in the medium to long term through the robust plan of the Nigerian Liquefied Natural Gas Limited (LNG) in collaboration with the Nigerian government to provide liquefied petroleum gas (LPG) through the National Gas Expansion program. This program seeks to expand domestic gas utilization as part of the National Gas Policy of government.

This program elevates LPG as the fuel of choice compared to other competing fuels and sets to establish a 20-million-cylinder injection scheme estimated to be between 5-10 million, introduced in a pilot project within the country. The government-backed its action with funding by providing an intervention fund called the LPG energy fund of 50 million dollars for the first phase in partnership with the African Export-Import Bank (Afrexim) for entrepreneurs and business start-ups to grow capacity. LPG is fast becoming a Nigerian choice product to drive gas optimization and utilization as a precursor for green hydrocarbon with zero-emission. The LPG utilization window becomes the forebearer to an eventual move toward bio-LPG. Adopting such frameworks of gas optimization enables other African countries, including Nigeria, to keep existing energy infrastructure in place, grow new capabilities and capitalize on the existing trajectory of progress in supply and distribution chains to reach net-zero at a far lower cost. Implementing such frameworks will, without doubt, help overcome the concern of stranded assets, stakeholders' fear of loss in investment, and consumers will not have to reinvest inexpensive electrical solutions if their gas appliances can be repurposed for zero-carbon LPG.

EMERGING TRENDS IN AFRICAN APPROACH TO ENERGY MIX AND TRANSITION

Various countries address the energy transition according to their peculiarities. Such peculiarities have situated the development of the resources, whether an oil-producing nation or not. For the oil-producing countries, energy transition is seen in three approaches.

1. The anti-fossil fuel approach - these are categories of African nations that embrace the anti-fossil fuel argument hook, line, and sinker. For them, the anti-fossil fuel group of the developed countries is legitimate and needs support to make a quick transition from fossil fuel to renewables. Such support has its incentive in the form of climate finance and technology transfer in foreign direct investment, with these incentives and assurances, they churn out policies that stifle investments in fossil fuels even with tens of billions of proven oil and gas reserves. They have started to distance themselves from fossil fuels. However, once a country starts distancing its economy from one of its mainstays, like oil in the first instance, it will move to gas soon. Africa's economic prosperity lies with the developed world for these counties, and Africa cannot make it without external support.
2. The rejectionist approach – countries in Africa under this category think energy transition is a conspiracy of the developed economies to stop the less developed but resource-endowed economies from making a significant leap into industrial development. Energy from fossil fuels is critical to the growth and survival of these countries. They posit that despite the established stance of the sciences behind climate change, they are not ready to abandon fossil fuels. Because such moves do not serve the long-term goal of economic prosperity until they achieve a socioeconomic development level that guarantees the basics of life for their population, these countries believe that Africa has contributed 2.5

-4% of global energy emissions. They see no justification for Africa to abandon its most affordable, available, and reliable energy source for an expensive source. More so when they believe that Africa is on the verge of an industrial breakthrough. For them, even when Africa has not been able to master the technology of the oil and gas industry despite decades of operation, Africa will continue to produce fossil fuels without external support. It will also compel them to develop regional and sub-regional energy infrastructure (Ihetu, 2021).

3. The third group is the pragmatist; these countries fall between the first and the second group. These countries agree that greenhouse emissions are harmful to the atmosphere and that every effort should reduce or end them. They also believe that the developed world championing it is futile for any developing or poor dependent country to go against it. However, they also recognize that they cannot hurriedly abandon fossil fuels. They affirm that they need a foreign market for their fossil fuels, even when they know that if they aggressively pursue their policy of creating a continent-wide market for the 1.3 billion inhabitants, nearly half of whom live without electricity or any form of modern energy. The total of oil and gas in Africa will not be enough for Africa.

Of these three groups highlighted, the third group (pragmatist) seems to be on the ascendancy in Africa today. This assertion is based on the consumption patterns of the elites within the continent. The elite wants to continue to get to the foreign market by exporting their oil and gas in their crude form to get the foreign earnings to continue importing luxury goods and services, but how long can we continue this way. African oil and gas producing countries cannot make progress until they can critically evaluate the dire situation; they will be with the successful pursuit of the Paris accord and Glasgow agreement on climate change.

OWNING ELECTRIC CARS IN AFRICA

In Nigeria, like most African countries, owning electric vehicles (EVs) seems like owning luxury items. The society is yet to accept the migration to these vehicles. In other jurisdictions like Canada, they have heightened the awareness of introducing an annual Electric Vehicle Road use tax (Hardman et al., 2017). The charge ranges from 100 to 150 Dollars to maintain roads and highways. Canada is finding additional funding to finance road maintenance. So, due to acceptance, discussing EV road tax and legislating renewable energy seems a viable option. It comes directly from taxpayers (EV owners) (Hardman et al., 2017; Jin & Slowik, 2017).

Although EVs are inexpensive to operate and maintain compared to fossil-fueled vehicles, irrespective of whether it is powered by propane or PMS. Buying an EV comes at a premium, and, for the most part, the owner tends to have higher-than-average incomes. Developed countries have therefore moved from accepting the concept of EVs to receiving significant tax payments from EV owners.

However, the situation in South Africa is different because EVs are part of the options of cars available. Brands such as BMW, Mini Cooper, Jaguar, Porsche, and Volvo are available. The problem, however, lies with the cost of owning the EVs. With a low adoption rate, making the transitioning to a zero-carbon future difficult for citizens, South Africa seems to be ahead of other African countries as the EVs entered the South African market in 2013 with inherent high purchase price, high battery price, and high likelihood for owning a secondary vehicle as an option. Another area of difficulty is limited charging stations around the country. This provides a business opportunity for entrepreneurs to increase charging points for EVs.

The South African case within the African context means that they are years ahead of other African counties. For the other countries with low socio-economic status, most of the people grapple with necessities of life being bordered by climate change issues. Interestingly green-tech start-ups are leading Africa's charge towards electric mobility across the continent (Dosso et al., 2021; Martins et al., 2021). There is an increased interest among developing countries in the potential of the African EVs market, and they are laying out comprehensive policy frameworks to catalyze the transition and adoption of electric mobility (Dosso et al., 2021; Martins et al., 2021).

Intriguingly, Africa has large Uranium deposits, a critical component for making EV batteries, in the Democratic Republic of Congo, Gabon, Namibia, Niger Republic, and South Africa (Ayers, 2020; DeBoom, 2020; Humphries, 2019). How this considerable potential will benefit the people of these countries is a different discussion (Ayers, 2020), but suffice it to say that we can grow the electric mobility industry and drive the production of EVs if we put our houses in order.

If harnessed, the electric mobility industry will provide opportunities for green jobs, especially in the electric two- and three-wheelers section of EVs. It will boost its potential for local manufacturing and assembly of EVs and business opportunities in charging and battery swapping. Countries like Kenya, Ethiopia, and Senegal are looking at EV uptake. Nigeria is more interested in moving from PMS to gas fuel vehicles which are not far-reaching in the energy transition. Additional issues surrounding electric cars are the stress on the electric grids. Electric vehicles require additional power be transmitted through the grid to serve as charging points at home, in offices, at the mall, and on the highways. We look forward seeing African countries adopt EVs and leverage new technology to grow society rather than adopt the wait-and-see approach. This has not helped in the 20th century and will not help in the 21st century. (Availability of electricity for everyday domestic use is still a huge challenge on the African continent. Electric vehicles will pose an additional challenge.)

EMBRACING LEADERSHIP OF ENERGY MIX ADOPTION

There is the need to move into unconventional forms of leadership as the conventional forms have failed us in many regards, or better still, the evils of capitalism far outweigh the good. This failure in leadership is not limited to a geographical location; instead, it is global as strikes, strives, wars, and rumors of wars directly result from the leadership failure across-board. Be it at the political level or organizational level.

There is the need to recognize that a democratic, market-based community-serving alternative in the choice of economic system is urgently needed, whether it is socialist, statist, or capitalist with their hybrids are welcomed. What bothers us are the rules they operate and how equitable are the local ownership of the means of production and resource allocation (Korten, 2016).

However, our focus is on the political and organizational leadership irrespective of the organization's industry and how they translate the economic system for the greater good of the people, leading to adopting a progressive, sustainable energy mix.

Every country or organization will not be able to adopt an energy mix at the same pace or in one swoop but embracing unconventional leadership that accepts business models which encourage reduced carbon footprint toward a net-zero target is desirable. Leaders who embrace change will benefit from a more innovative work environment that allows employees to work remotely, onsite, or hybrid, depending on the peculiarities of the organization. Therefore, organizations should continue to embrace technology and give employees better chances of improving productivity and the overall success of the organization.

CONCLUSION

In this chapter, we have established the devastating effects of climate change on African countries. However, the general conversation on Africa tends to focus more on the physical risks. However, the climate models show that the continent is among the world's regions likely to be hardest hit by the changing climate. A substantial percentage of the continent's population is without access to energy, living in extreme poverty, and has low levels of resilience and adaptation. These conditions are already complex and dire but might exacerbate the socioeconomic impacts of climate change.

Given this situation, one is forced to ask why African nations are among the lightest polluters yet suffer the most from climate change in a bid to have water, food security, sustainable development, and economic sustainability?

While we cannot exhaustively answer this and other questions, governance and politics must encourage a Green Africa. We could do our bit to drive improvements in energy efficiency and vehicle fuel economy, increase wind and solar power, increase biofuels from organic waste (Scarlat et al., 2018), set a carbon price, and protect forests. These measures are practical ways to reduce the amount of carbon dioxide in the atmosphere.

Climate change might also mean increasing the capacity to improve renewable energy sources. It will mean a reduction in fossil fuel energy sources, thereby shifting global demand patterns away from fossil fuels and most African countries' commodities exports come from fossil fuels. This might have an immediate and long-term impact on revenue earnings for development initiatives leading to a reduction in gross earnings, but the benefit of a sustainable planet to is more beneficial than increasing earnings because we can diversify our revenue portfolio.

What we are witnessing presently are many global initiatives are working towards resolving the energy crisis and have introduced regulations and restrictions on carbon emissions, promoting greener manufacturing and construction projects, funding research into hybrid technologies and more sustainable technologies, and more. Provinces, counties, and cities see beyond the recycle bin and recognize that using their local resources is essential. More community gardens, parks, and farmer's markets are springing up to introduce more sustainable elements into people's lives and as an essential part of educating the public about the importance of resources.

REFERENCES

Aceto, G., Persico, V. & Pescapé, A. (2018). The role of Information and Communication Technologies in healthcare: taxonomies, perspectives, and challenges. *Journal of Network and Computer Applications.* 107 (1) 125-154. https://doi.org/10.1016/j.jnca.2018.02.008

Ayers, W. N. (2020, September). Cradle to Grave Issues with Vessel Lithium-Ion Batteries. In *SNAME Maritime Convention.* OnePetro.

Buhaug, H., Gleditsch, N. P., & Theisen, O. M. (2008). Implications of climate change for armed conflict. *Washington, DC: World Bank.*

DeBoom, M. J. (2020). Climate necropolitics: ecological civilization and the distributive geographies of extractive violence in the Anthropocene. *Annals of the American Association of Geographers, 111*(3), 900-912. https://doi.org/10.1080/24694452.2020.1843995

Demırbas, A. (2017). The social, economic, and environmental importance of biofuels in the future. *Energy Sources, Part B: Economics, Planning, and Policy, 12*(1), 47-55. https://doi.org/10.1080/15567249.2014.966926

Dosso, M., Nwankwo, C. I., & Travaly, Y. (2021). The Readiness of Innovation Systems for the Fourth Industrial Revolution (4IR) in Sub-Saharan Africa. In *Entrepreneurship, Technology Commercialisation, and Innovation Policy in Africa* (pp. 13-32). Springer, Cham.

Ganivet, E. (2020). Growth in human population and consumption both need to be addressed to reach an ecologically sustainable future. *Environment, Development and Sustainability, 22*(6), 4979-4998.

Gilbert, A., Bazilian, M. D., & Gross, S. (2021). The Emerging Global Natural GasA Market and the Energy Crisis of 2021-2022.

Hardman, S., Chandan, A., Tal, G., & Turrentine, T. (2017). The effectiveness of financial purchase incentives for battery electric vehicles–A review of the evidence. *Renewable and Sustainable Energy Reviews, 80*, 1100-1111. https://doi.org/10.1016/j.rser.2017.05.255

Humphries, M. (2019). Critical Minerals and US Public Policy. *Congressional Research Service: Washington, DC, USA.*

Ihetu, G. (2021). Still on renewables and the future of Gas. Thisday Newspapers. https://www.thisdaylive.com/index.php/2021/09/27/still-on-renewables-and-the-future-of-gas/ September 27.

Jin, L., & Slowik, P. (2017). Literature review of electric vehicle consumer awareness and outreach activities. *International Council on Clean Transportation. Available from internet:* https://www. *theicct.org/sites/default/files/publications/Consumer-EV-Awareness_ICCT_Working-Paper_23032017_vF. pdf.*

Martins, R. M., Park, E., Hain, D. S., & Jurowetzki, R. (2021). Mapping Entrepreneurial Ecosystem for Technology Start-ups in Developing Economies: An Empirical Analysis of Twitter Networks Between Start-ups and Support Organizations of Nairobi's Digital Economy. In *Entrepreneurship, Technology Commercialization, and Innovation Policy in Africa* (pp. 55-97). Springer, Cham.

Moeletsi, M.E. Socio-Economic Barriers to Adoption of Electric Vehicles in South Africa: Case Study of the Gauteng Province. World Electr. Veh. J. 2021, 12, 167. https://doi.org/10.3390/wevj1204016

Nadel, S., & Ungar, L. (2019). Halfway there: Energy efficiency can cut energy use and greenhouse gas emissions in half by 2050. *Report u1907 American Council for an Energy-Efficient Economy (ACEEE).*

Newbery, D. (2016). Missing money and missing markets: Reliability, capacity auctions and interconnectors. *Energy policy, 94*, 401-410.

Onukwue, A. (2022). Stressful Times.: The price of diesel in Nigeria has doubled. *Quartz Africa.* https://qz.com/africa/2142918/the-price-of-diesel-in-nigeria-has-doubled/

Skea, J. I. M., & Nishioka, S. (2008). Policies and practices for a low-carbon society. *Climate policy*, *8*(sup1), S5-S16. https://doi.org/10.3763/cpol.2008.0487

Statistics Canada. Table 18-10-0001-01 Monthly average retail prices for gasoline and fuel oil, by geography. DOI: https://doi.org/10.25318/1810000101-eng

Tverberg, G. (2021). The Energy Crisis That No One Is Talking About. *Oil Price*. https://oilprice.com/Energy/Energy-General/The-Energy-Crisis-That-No-One-Is-Talking-About.html

UNICEF (2020). Massive flooding in Mozambique, Malawi, and Zimbabwe. *UNICEF bringing relief for children and families affected by Cyclone Idai and Cyclone Kenneth* https://www.unicef.org/stories/massive-flooding-malawi-mozambique-and-zimbabwe

World Energy & Climate Statistics – Yearbook 2021. https://yearbook.enerdata.net/renewables/wind-solar-share-electricity-production.html

Zappa, W., Junginger, M., & van den Broek, M. (2021). Can liberalized electricity markets support decarbonized portfolios in line with the Paris Agreement? A case study of Central Western Europe. *Energy Policy*, *149*, 111987. https://african.business/2021/09/technology-information/opportunities-for-africa-in-electric-vehicle-market/

Chapter 11

Managing The Dynamics

In Individual and Government

By Sheri Harshaw & Compreca C. Martin

INTRODUCTION

The United States of America (USA) is a country other nations look up to for resources and partnership opportunities. It is one of the largest continents compared to Asia, Africa, and Europe (World Population Review, 2022). The USA was incorporated in London in 1783 and was a territory of Great Britain (National Liberty Health Alliance, n.d.). Before 1776, Great Britain had full ownership of the USA. According to research, the colonist did not win the revolutionary war, which means the USA was still owned by Great Britain (Mount Vernon, n.d; National Liberty Health Alliance, n.d.). Troops did not leave until after 1796 (National Liberty Health Alliance, n.d.).

The war started because of intolerable acts, tax issues, the Boston Tea Party, and King George's III parliament speech (Tomes, 2021). King George III of England financed both sides of the revolutionary war (National Liberty Health Alliance, n.d.). After the war, the USA became a corporation and started creating policies and procedures for American citizens to follow (Freeman, 1991). One of the first policies was the Articles of Confederation and the U.S. Constitution, which meant the start of a new nation (Mount Vernon, n.d).

With the help of James Madison, Henry Knox, and John Jay, they could all able to give advice and assist with bringing the policy to fruition (Mount Vernon, n.d.). Each envisioned a national government with separate legislative, judicial, and executive branches with one central government to control the American people (Mount Vernon, n.d.). Once these new laws were created, they left many Americans fearful, overwhelmed, and with limited understanding (Freeman, 1991).

This chapter provides an understanding of some of the dynamics, such as the Individual perspective, the way the government is structured, the student loan process, cultural diversity, the health care infrastructure, international aid, terrorism, currency, and taxes. This chapter will end with a summary.

THE DYNAMICS OF INDIVIDUAL

The dynamics of an individual starts at conception and matures into personality traits imprinted on the child by the parents and supporting village. Therefore, at birth, a child will start exhibiting personality traits cultivated by environment, and family, identified as the beginning phase of the dynamics of an individual (Fryxell & Lundberg, 2013).

Case studies have shown that the dynamics of an individual is a compression of events that evolved from evolutionary and population ecology that generalize the individual interactional behavior. Kareiva 1990: Gilpin and Hanski, 1991 note that the areas of individual dynamics based on the metapopulation theory describe the evolutionary framework. That references behavioral patterns (i.e., patterns exhibited by emigration probability). That defines the frequency and density of individual behavior.

The purpose of this subchapter is to review multiple case studies and gather information about the dynamics through the individual through the recognition and demonstration of their individuality based on cultural traits, personal beliefs, and environmental impact. The dynamics of the individual can impact the daily fiber of effective functionality in society based on individualized beliefs and not singular traits. Therefore, multiple case studies have provided different views, allowing the reader(s) to draw their thoughts and conclusions (Fryxell & Lundberg, 2013).

The problem is the dynamics of an individual motivated by personal beliefs and not traits which should recognize individual differences, creating a versatile society, which should respect and accept without bias and discrimination.

Cultural Traits

Cultural traits are also referenced as embedded behavior based on identifying cultural traits similar to human cultural evolution which emphasizes the different underlying dynamics of an individual (Mesoudi, 2021). The underlying human traits originate from cultural selection, which has identified cultural variants (i.e., single-parent households, an only child, and race). These variants can impact behavioral patterns (i.e., personal traits) relevant to the dynamics of individual growth and maturity. Cultural traits support the dynamic of individuals as an invisible thread holding together the individual uniqueness, which makes everyone different.

Personal Beliefs

Personal beliefs characterize the dynamics of an individual, which distinguishes personal characteristics that have shaped great leaders, scholars, and practitioners. Personality beliefs or traits can be categorized as a popular conceptional framework described by five domains: Neuroticism, Extraversion, Openness to Experience, Agreeableness, and Conscientiousness (McCrae & Costa, 2003).

Neuroticism is the core identifier of how an individual reacts to negative and positive situations exhibited in everyday life (Hisler et al., 2020). Therefore, the experience and handling of mood-altering incidents impact the dynamics of an individual applicable to the psychopathology of neuroticism (Widiger, 2011). Whereby, higher levels of conscientiousness predict longevity (Kern et al., 2009). Agreeableness is related to higher satisfaction with relationships (Malouff et al., 2010). Of relevance, research has demonstrated that personal beliefs vary based on gender, age, and culture. However findings are mixed and may be contingent based on the level of personality scales used, based on culture (Kaiser et al., 2020).

Environmental Impact on Dynamics of Individual

The dynamics of individual surroundings is core to personal success, how special circumstances are handled, and the judgement of situations. The environmental impact is the structural-based typology with representative samples based on the specific population (i.e., the younger generation, the older generation, and other cultures (Maya-Jariego, 2021). Maya-Jariego and Holgada, 2015 analyzed data from 403 personal networks and other studies on metropolitan life cycles: the significance of the environmental impact on different generations. The new idea level of thinking and internet interaction used by the younger generation. The older generation still relies more on face-to-face interaction as a form of networking.

Environmental impact relevant to different cultural reasoning and understanding have become a loud voice pertinent to individual differences and beliefs in relation to the dynamics of individual. Therefore, all members of the network are not held in equal regards, but instead of human social networks and the layered structure of who they know as a network of succession (Dunbar & Spoors, 1995; Hill & Dunbar, 2003, Zhou et al., 2005)

MANAGING THE DYNAMICS IN GOVERNMENT

Dynamics in Government Structure

There are constant changes in how the U.S. government implements models and policies for citizens. With constant changes, citizens may lack knowledge about some of the policies and models that are approved (Garman, Johnson, & Royer, 2011). The USA government is set up based on several different legislatures of power, such as legislative, judicial, and executive branches. This setup is unique compared to other countries where there is only one authoritative figure making all the decisions. In the U.S. power is divided between federal and state to create an equal balance. The role of the federal government is to provide funding and policy updates to states. The state is responsible for ensuring that the funding is used to create programs resources, and encouragement of economic growth in communities.

The federal government, there is a student loan agreement between for-profit colleges, universities, and banks to offer student loans that has nothing to do with the state.

Student Loans

The student loan program is the second leading cause of debt in the U.S. (Shy, 2022). It accounts for more than a trillion dollars of outstanding debt from more than 40 million borrowers (Shy, 2022). Many for-profit colleges and universities in the U.S. have been investigated for fraud (Kutz, 2010). The investigation started because of a significant increase in enrollments. For-profit colleges grew from about 360,000 students to more than 1.8 million in the last several years (Kutz, 2010). The student loan environment, mostly determined by federal policies, affects a student's welfare after completing high school (Card & Solis, 2022). Institutional grants and Federal Pell grants are the largest sources of financial aid (Biswas, n.d.).

Consumers qualify for these grants by completing a Free Application for Federal Student Aid, also known as (FAFSA). After receiving FAFSA, the office of Federal Student Aid uses a legally defined formula to calculate a student's expected family contribution (EFC) based on family size and income criteria.

The gap between an institution's cost of attendance (tuition, fees, and room and board, net of any scholarships or grants) and the student's EFC defines the financial need (Biswas, n.d) Only students who have a positive financial need may borrow subsidized federal loans, where the federal government pays for the interest accrued during enrollment (Biswas, n.d.).

Loan availability and loan terms inform the decision to borrow, which in turn finances the type of investments the student makes to develop human capital. The problem is that students graduate from post-secondary education and do not get the salary they hoped for (Shy, 2022). Because of this concern, President Joe Biden announced his decision in 2022 to review Federal student loan policies (CNBC, 2022). For many borrowers, this could mean complete student loan forgiveness or a reduction in the amount to be paid back (Federal Reserve, 2022).

One of Biden's goals is to cancel $10,000 for all borrowers who earn less than $150,000 per year (CNBC, 2022). The issue remaining is that many students have an average debt balance of $50,000-$250,000, which means they may remain in default or delinquent despite Biden's efforts (Card & Solis, 2022). Canceling $10,000 per borrower would forgive a total of $321 billion of federal student loans and eliminate the entire balance for about 11.8 million borrowers (CNBC, 2022). The average borrower would receive $8,478 in student loan forgiveness (Federal Reserve, 2022).

In comparison to other countries such as the United Kingdom (UK) student loan programs are more reasonable. Money for research is allocated through the Higher Education Funding Council of England (HEFCE). HEFCE is responsible for allocating funding for larger universities and higher education institutions (Promades, 2012). HEFCE distributes money to institutions based on a number of factors, such as types of research undertaken, number of students, demographics. Once these factors are considered the HEFCE distributes a block of money in the form of a grant, which allows the institution to spend the money in areas that are of high importance (Promades, 2012). High importance areas may consist of research, teaching resources, etc. The UK pays for post-secondary education, relieving the burden and stress from students and families (Promades, 2012). The average spending on education is around 90 billion annually (Promades, 2012).

The U.S. could benefit from revising the student loan policy by identifying what other international markets are doing to provide access to education. There is an increased need for cultural diversity in the U.S., which will require access to education. Cultural diversity is considered at both the state and federal levels.

CULTURAL DIVERSITY

Culture is referred to as patterns of shared meanings (Sanger, 2008). According to Sanger, an organization's culture has been compared to a person's personality and character. An innovative program in any organization must be aligned with a company's culture (Rosenfeld, 2012). Before 1848, there was no diversity in the USA federal government (Black Americans in Congress, 2007). Fredrick Douglas became one of most important leaders of the movement for African-American civil rights but experienced a significant amount of grief because slavery was still legal during that time (Black Americans in Congress, 2007).

After Fredrick Douglas, minorities would run for candidacy but not get elected (Black Americans in Congress, 2007). There was no change in cultural diversity until 2008, when President Barack Obama was voted into office as the 44th President and during that time, slavery was illegal (White House.gov, n.d)

According to Avolio and Yammarino (2008), patterns of leadership behaviors are likely to change by 2034. Women and different ethnicities will transition into leadership roles, which will require more training. In 2034, empowerment, collaboration, and continuing change in an organization will ingrain in values and beliefs (Avolio & Yammarino, 2008). Part of transforming requires being able to work in a diverse workplace (Avolio & Yammarino, 2008). Rosenfeld (2012) stated innovation preferences of its people are appropriate for innovation and organizational goals.

Leaders should be carefully selected based on what the organization needs to help them grow, not based on gender or nationality. In 2021, President Biden was appointed and signed into the legislature as the first African American Vice President and African American supreme court judge (White House gov n.d). The appointment gave American citizens hope that in the future, there could potentially be different nationalities and genders running for presidency and federally appointed positions (Chisholm-Burns et al., 2017).

At least 25-30 percent of women are in leadership roles ranging from the lower, middle, and executive positions across all industries (Chisholm-Burns et al., 2017). With this significance, it is evident that in the near future more women of different ethnicities and nationalities could be appointed to federal government and health care positions. The healthcare industry in the U.S. will see an increase of about 50% minorities by 2024 (Gupta, 2019).

HEALTHCARE INFRASTRUCTURE

Healthcare accounts for one-sixth of the US economy, which makes it the largest industry (Folland, Goodman, & Stano, 2010). Healthcare is different from other economic fields because it is comprised of several markets and systems. Each market has a complex team with morals and ethics to provide a high-quality service to patients to receive money. In healthcare the type of quality provided depends on making a life-or-death decision for an individual's health and welfare.

One of the theories applied in healthcare is welfare. Welfare is where individuals use their best judgment on what is best. This information is applied

in healthcare to determine what is in demand. For example, patients have the option to choose providers based on cost, services, and products.
There are two core concepts of financial analysis when looking at spending, and those are timing and risk (Getzer,, 2007).

Looking carefully at how people make deals with physicians, hospitals, and with each other is a fundamental approach to health economics (Getzer, 2007). A review of the healthcare market is essential in determining what the needs and wants are for the U.S. population. Every time a new President takes office, there is a discussion about changes to health care policies. Constant changes occur because of the structuring. There are several cost containment initiatives, such as Medicare and Medicaid, Affordable Care Act (ACA), and Managed Care Organizations (MCOs), and escalating costs, which cause an ongoing debate.

Managed Care Organizations

A Managed Care Organization (MCO) is a single organization at the state level which manages the financing of insurance to provide health care services (Getzer, 2007). Financing plans under MCO are negotiated between the employer or individual. The negotiated contract will include a fixed premium per enrollee and the health services provided. Managed care ranks as the lower-cost option insurance that patients can purchase, and many may choose incorrect plans due to limited understanding (Getzer, 2007). Patients are not educated about choosing the insurance packages necessary to meet their needs (Getzer, 2007).

Patients rely on health care coverage to pay for services, such as heart transplant. The leading cause of death in the US is heart disease, which increases intensity and implication costs (Greenwald, 2010). MCOs will pay a percentage of procedures, leaving a patient to pay for the rest. Managed care organizations place a cap on the number of services they are willing to pay for. Based on findings from Folland, (2010) healthcare organizations have complained that MCOs underprovided services, which may be harmful to patients. Most healthcare organizations in the U.S. rely solely on Medicare and Medicaid reimbursements (CMS.gov, 2022).

Medicare and Medicaid

Medicare and Medicaid are the largest federal reimbursement programs in the U.S. (CMS.gov, 2022). Medicare is available to people aged 65 or older, younger people with disabilities and people with End Stage Renal Disease (Medicare.gov, 2022). There are four parts covered under Medicare which are Part A (Hospital Insurance), Part B (Medicare Insurance), Part C (Secondary), and Part D (Prescription drugs) (Medicare.gov, 2022).

One of the ways Medicare controls the reporting process is through documentation. Third parties rely on shared information to monitor care received by constituents and for purposes of promoting accountability. A third party may be a government agency, such as Medicare, to generate and disseminate information to consumers to guide their purchasing decisions and improve access to preventive care.

Medicaid provides health coverage to millions of Americans, including eligible low-income adults, children, pregnant women, elderly adults, and people with disabilities (Medicaid.gov, 2022). Medicaid is administered by states, according to federal requirements. The program is funded jointly by states and the federal government (Medicaid.gov, 2022). Another policy, called the Affordable Care Act (ACA), was enacted in 2010 (Healthcare.gov, 2022).

Affordable Care Act

Under the ACA, all American citizens could have access to healthcare by completing an application and qualifying for a plan based on income, family size, and state (Healthcare.gov, 2022). The underlying dynamic with the ACA is that with other similar changes to the healthcare system will not necessarily solve all access to healthcare issues (Kenney et al. 2012).

Sanders (2013) contends that the ACA was deemed constitutional by a divided Supreme Court in June 2012, which required all residents of the U.S. to retain minimum essential health coverage and imposed a tax penalty for failure to do so. According to Sanders (2013), the individual mandate is softened by tax credits or subsidies, made available to eligible individuals purchasing insurance through one-stop-shop exchanges.

According to Sanders (2013), under the Commerce Clause of the US Constitution, the federal government holds some authority to regulate insurance. However, it is an area in which federal and state powers overlap. Sanders (2013) emphasizes that early drafts of the ACA envisioned those states would establish exchanges under section 1311(b) (1). The addition of federal exchanges as a safety net changed state exchanges from a requirement to an option, restoring the provision's constitutionality (Sanders, 2013).

Costs

The U.S. appears less favorable when it comes to healthcare in comparison to other countries (Nayyer, 2011). According to Kent (2008), prescription drug prices are about 35% to 55% higher in the U.S. than they are in other countries. Spending on prescription drugs increases annually compared to inflation rates (Feng, 2009). According to Greenwald (2010), health care costs are the most expensive in the U.S. compared to other countries. Despite the increase in costs, the U.S. Federal government continues to lend a helping hand to foreign countries (Jadallah, 2014). Sending billions of dollars to foreign countries in need limits access to care for U.S. citizens (Johnson & Stoskopf, 2010).

INTERNATIONAL AID

The U.S. foreign policy has three interlinked targets for entering countries through free markets, democracy, and stability, which are beneficial for the world (Jadallah, 2014). U.S. officials are known to send aid to other countries, but some countries are not in agreement with U.S. policies (Jadallah, 2014). This signifies that the US can lend some solutions, but not to all countries. There are other countries, such as Japan, the United Kingdom, and Canada that have a more attractive health care system (Johnson & Stoskopf, 2010).

There are no approximate numbers of how many Americans travel to other countries for care; however, the National Coalition of Health listed a statistic of 500,000 (Garman, Johnson, & Royer, 2011). American consumers travel to other countries that have cheaper rates and have stable health care model and policies (Johnson & Stoskopf, 2010). Johnson and Stoskopf shared detailed information about countries and how they may or may not pose a threat to the U.S. With international aid and people traveling across countries there remains a concern for terrorism.

TERRORISM

Political violence and terrorism pose a unique challenge in the U.S compared to transnational foreign terrorist (Shaffer, 2022). Almost every year there is an attack in the U.S. that leave citizens in fear (Shaffer, 2022). Terrorist attacks are identified in three categories: Jihad terrorism, racial terrorism, and political terrorism.

Jihad Terrorism is a foreign terrorist organization. Under this organization the following attacks occurred, the Times Square bombing, Boston Marathon attack, Orlando nightclub shooting, New York City truck attack, and Naval Air Station attacks (Phares, 2014). The U.S. army successfully captured Al Qaeda, who was responsible for most of these attacks (Phares, 2014). Despite the success in capturing Al Qaeda, there is still an ongoing concern in regard to racial terrorism attacks.

Over recent years, violent incidents have often been underpinned by *racial, ethnic, political, and ideological* motivations (Yar, 2022). Some racial attacks that have occurred were the Wisconsin Sikh Temple shooting, the Charlotteville car attack, the Pittsburgh Synagogue shooting, and El Paso shooting (Yar, 2022). These are not foreign concerns but rather a country issue. The Department of Justice (2012) lists the following statistics, 25% of victims were Caucasian, 25% were Hispanic, 40% were Black, and 10% were of other races. According to Yar (2022), racial terrorism is something that must be handled at both the state and federal levels.

Political terrorism has been defined as a politically motivated assassination, arson, kidnapping, and other forms of violence (Shaffer, 2022). Some attacks in the U.S. were the 9/11 attack, anthrax, the assassination of George Tiller, the congressional baseball shooting, and the U.S. mall bombing (Shaffer, 2022). In 2003, President Bush discovered there was a linking between terrorism and sex-trafficking (Hesford, 2019). The discovery of global sex-trafficking led to more surveillance and strict border protections in the U.S. (Hesford, 2019).

Texas and Washington were the first states to report human trafficking (Santana, 2018). In 2017, Dallas, Texas, had more than a hundred sex-trafficking cases (Santana, 2018). Texas and California have the highest hub for sex trafficking (Santana, 2018). From 2017-2019 San Francisco, California, had 70%, and Portland, Oregon, had 85% reported sex trafficking cases (Santana, 2018).

With an increased concern for terrorism, the U.S. is now worried about mass shootings occurring in public places (National Institute of Justice, 2019). There have been a significant amount of mass shootings occurring in places such as schools, churches, parks, and movie theaters (National Institute of Justice, 2019). The federal and state legislatures are collaborating to revamp gun laws and policies to get this under control (National Institute of Justice, 2019). Aside from reviewing the gun laws, the Federal government must also review U.S. Currency. There is research referencing the U.S. currency as fake or false dollars (National Liberty Health Alliance, n.d.).

U.S. CURRENCY

The currency system in the US is a dollar currency. Every country has a different currency, and the U.S. currency may not be stable (Getzer, 2007; National Liberty Health Alliance, n.d.). The American citizen is put in a position to trust that the currency is real based on the saying. "In God we Trust.." However, many people debate about whether there is a God and how did this all come about. Many U.S. citizens question whether the

American currency is real money, is debt real, and if there is such thing as a mortgage.

For more than a hundred years, most currencies have been issued by central banks, while except for cash, retail payment systems have typically been implemented by the private sector (Wang, 2022). The Federal Reserve was created in 1913 by the Federal Reserve Act to serve as the nation's central bank (National Liberty Health Alliance, n.d.). The Board of Governors in Washington, D.C. is an agency of the federal government and reports to and is directly accountable to Congress (Wang, 2022). The Federal Reserve bank is a private company setup, and as is every court and every police force and even Congress is a company and not a person (National Liberty Health Alliance, n.d.). The purpose of any commercial for-profit company or corporation is to make money for its owners. The government would not do anything that earns money, but rather they take money from citizens (National Liberty Health Alliance, 2021).

According to Freeman (1991), real currency is only in gold or silver. In 1945, Britain, with an agreement, established the dollar as the world's currency, which meant that international commodities were priced in dollars. Under this agreement, the U.S. had an advantage under the condition that those dollars would remain redeemable for gold at a consistent rate of $35 per ounce (Freeman,1991). The U.S. promised not to print much money but later did because the Federal Reserve refused to allow any audit or supervision of its printing processes (Freeman, 1991).

The U.S. started printing far more money than it had in gold, which led to a rapid decline in the value of the dollar (Freeman, 1991). Once this happened, the U.S. began to accumulate debt. In 1973, President Nixon asked King Faisal of Saudi Arabia to accept U.S. dollars in payment for oil and to invest any excess profits in U.S. Treasury Bonds and notes and bills (Freeman, 1991). It led to the exchange of goods for federal service notes. Thus, in return, allowing the U.S. to maintain a constant supply of federal reserve paper.

Those who control the U.S. understand that if even a few countries begin to sell their oil in another currency, it will set off a chain reaction, and the U.S. dollar will collapse (Freeman, 1991). The issue remains in public policy versus public law in America. America is a bankrupt nation owned by creditors which are the countries that agreed to accept federal reserve notes for oil (National Liberty Health Alliance, n.d.). With high debt accumulating, the American citizen is forced to work and pay taxes (Sanders, 2013).

Taxes

The National Liberty Health Alliance (n.d.) explained the government's main goal is for U.S. citizens to work hard and pay them money. Each American citizen has a strawman account set up under a fictitious name which will help pay the burden of debt the U.S. has encountered (Freeman, 1991). The government and police officers uphold the law, and taxes are essential in order to keep things going ((National Liberty Health Alliance, n.d.; Sanders, 2013). Under strict statutes, American citizens are told they must abide by the law of paying taxes if they work in the U.S. (Sanders, 2013). Taxes are controlled through a third-party agency called the Internal Revenue Service (IRS).

The IRS is a third-party private corporation hired by the federal government to oversee the payment of taxes (National Liberty Health Alliance, n.d.). The average tax debt in the U.S. is approximately $17,000 (Sanders, 2013), and the IRS reported there were approximately 14 million delinquent taxpayer accounts by the end of 2019 (IRS.gov, 2019).

Failure to pay taxes could lead to federal tax court, citizens turning over their passports, losing their homes, and even garnishment of wages (IRS.go v, 2019). The problem is that the average U.S. citizen lacks knowledge about the IRS process (Sanders, 2013). Sovereign citizens do not believe in paying taxes and have accused the federal government of committing fraud by hiring the IRS (National Liberty Health Alliance, n.d.). If the IRS is not a legal entity, how do they have the power to take away a person's assets?

According to Freeman (1991), if the IRS sends a notice of deficiency, even if it is for ten million dollars, under the common law, citizens are not required to file or pay taxes. Never argue the amount or facts in a tax case, seek solutions from the problem resolution officer or district director at the IRS tax agency (Freeman, 1991). IRS agents are hired to write delinquent and collection letters on behalf of the district director (National Liberty Health Alliance, n.d.).

In 1980 Freeman (1991) was successful in contacting the district director. Freeman said to the director, "I thought you ought to know there are agents working for you who are writing letters over your name with which you wouldn't agree. Do you read all the mail that goes out of this office over your signature?" (p. 5). The director responded, "Oh, I couldn't possibly read everything as it goes out by the bag full." (p.5). Freeman then responded, "the letters contradict the decisions of the Supreme court of the U.S. and the IRS is not doing it over their name, they are doing it over your name" (p.5). The director reviewed the documentation, and Freeman have not filed taxes since 1969 (Freeman, 1991).

The position of this chapter does not serve as a legal advice or stating that citizens should not file taxes. It is rather for education purposes only. The evidences shared might help individuals understand the dynamics within the U.S. government. I recommend a review of the common law book and UCC codes which can normally be purchased on Amazon or looked at in the library.

CONCLUSION

In summary, although several foreign countries admire the U.S., there is a considerable number of dynamics, such as the overall structure of the government, significant student loan debt, increased need for cultural diversity, health care infrastructure, international aid concerns, terrorism attacks, currency, and payment of taxes.

The dynamics of the individual is a defining factor that imprints life identifiers from conception and transposes through recognizable personality traits. These personality traits are mostly inherited from parents but mainly learned from life-learned experiences. These life-learned experiences: cultural traits, personal beliefs, and environmental impact are lateral differences between individualism and reductionism.

The reductionism of the dynamics of individuals removes the natural aspect of human behavior and polarizing behavior as can inability within our social setting. Therefore, providing social insight that is capitalized by the young generation, the older generation, and cultural differences as it relates to the overview of the dynamics of individuals, which is the most significant visual of our differences.

REFERENCES

Avolio, B., & Yammarino, F. (2008). *Transformational and Charismatic Leadership: The Road Ahead* . WA, UK: Emerald Group Publishing Limited.

Black Americans in Congress (2007). Retrieved from: https://www.govinfo.gov/content/pkg/GPO-CDOC-108hdoc224/pdf/GPO-CDOC-108hdoc224.pdf

Biswas, S. The Life-Cycle Impacts of Federal Student Loans: A Brief Literature Review. https://www.philadelphiafed.org/-/media/frbp/assets/institutional/banking/surf/spotlights/2022/surf_2022_q2-spotlight.pdf

Card, D., & Solis, A. (2022). Measuring the effect of student loans on college persistence. *Education Finance and Policy*, *17*(2), 335-366.

Chisholm-Burns, M. A., Spivey, C. A., Hagemann, T., & Josephson, M. A. (2017). Women in leadership and the bewildering glass ceiling. *American Journal of*

Health-System Pharmacy, 74(5), 312–324. https://doi-org.ezp.waldenulibrary.org/10.2146/ajhp160930

CMS.gov (2022). Medicare and Medicaid. Retrieved from: https://www.cms.gov/Medicare/Medicare

CNBC. (2022). President Biden's decision on student loan forgiveness. Retrieved from: https://www.cnbc.com/2022/06/23/what-bidens-student-loan-forgiveness-decision-will-mean-for-borrowers.html

Dunbar R., and Spoors, M., (1995). Social networks, support cliques, and kinship. Human Nature, Vol. 6., pp. 273-290

Elziadi, M., & Qassis, F. (2020). *Cultural Intelligence (CQ): Framing the Effectiveness of Leader-Follower Relationship* (Dissertation). Retrieved from http://urn.kb.se/resolve?urn=urn:nbn:se:lnu:diva-95541

Federal Reserve. (2022). Who are the federal student loan borrowers and who benefits from forgiveness? Retrieved from: https://libertystreeteconomics.newyorkfed.org/2022/04/who-are-the-federal-student-loan-borrowers-and-who-benefits-from-forgiveness/

Feng, Y. (2009). *Medicare prescription drug legislation: An analysis of economic effect.* (Order No. 3381608, University of Pennsylvania). *ProQuest Dissertations and Theses,* , 129-n/a. Retrieved from http://search.proquest.com/docview/304977903?accountid=35812. (304977903).

Folland, Goodman, &Stano (2010). The economics of health and healthcare (6th ed.) Upper Saddle River, NJ: Prentice Hall.

Freeman, H. (1991). The UCC connection how to free yourself from legal tyranny. National Health Alliance. Org

Garman, A.N., Johnson, T.J., & Royer, T.C. (2011). *The Future of Healthcare: Global Trends Worth Watching*. Retrieved from The University of Phoenix eBook Collection.

Getzer (2007). Health economics and financing (3rd ed.). Hoboken, NJ: John Wiley & Sons, Inc.

Gibb, J. R. (1967). Dynamics of leadership. UMHE Publications. Louis, Missouri.

Integrity Solutions LLC, (2017). Coaching and training for management: How to Be an Effective Leader. Vol.2., Pp.1-65.

Greenwald, H. P. (2010). Health care in the United States: Organization, management, and policy. San Francisco, CA: Jossey-Bass.

Gupta, A. (2019). Women leaders and organizational diversity: their critical role in promoting diversity in organizations. *Development & Learning in Organizations, 33*(2), 8–11

Healthcare.gov (2022). The Patient Care and Affordable Care Act. https://www.healthcare.gov/

Hesford, W. S. (2019). Trafficking American Exceptionality, Women's Studies in Communication, DOI: 10.1080/07491409.2018.1544007

Hisler, G. C., Krian, Z., DeHart, T., and Wright, A. G. C. (2020). Neuroticism as the intensity, reactivity, and variability in day-to-day affect. Journal of Research in Personality, 87(2020), p. 103964, 10.1016/jrp.2020.103964

IRS.gov (2019). Individuals with significant tax debt should act promptly to avoid revocation of passports. Retrieved from: https://www.irs.gov/newsroom/individuals-with-significant-tax-debt-should-act-promptly-to-avoid-revocation-of-passports

Kai N. Klasmeier, Jens Rowold, (2021). A diary study on shared leadership, teamwork engagement, and goal attainment, Journal of Occupational and Organizational Psychology, 10.1111/joop.12371, 95(1), Pp. 36-59.

Kaiser, T., Del Giudice, M. and Booth, T., (2020). Global sex differences in personality: Replication with an open online dataset. Journal of Personality, 88 (3) (2020), pp. 415-429, 10.111/jopy.12500

Jadallah, D. (2014). *United states economic aid: Imperfect hegemony in Egypt.* (Order No. 3614019, The University of Arizona). *ProQuest Dissertations and Theses,* , 408. Retrieved from http://search.proquest.com/docview/1513234336?accountid=35812. (1513234336).

Johnson, J.A., & Stoskopf, C.H. (2010). *Comparative Health Systems: Global Prospectives.* Retrieved from The University of Phoenix eBook Collection.

Kenney, G. M., McMorrow, S., Zuckerman, S., & Goin, D. E., (2012). A decade of health care access declines for adults holds implications for changes in the affordable care act. *Health Affairs, 31*(5), 899-908. Retrieved from http://search.proquest.com/docview/1015200874?accountid=35812

Kent, S. A. (2008). *Is more necessarily better? the impact of prescription drug coverage on clinical outcomes in a Medicare population.* (Order No. 3328598, University of Pennsylvania). *ProQuest Dissertations and Theses,* , 125-n/a. Retrieved from http://search.proquest.com/docview/304511137?accountid=35812. (304511137).

Kutz, G. D. (2010). For-Profit Colleges: Undercover Testing Finds Colleges Encouraged Fraud and Engaged in Deceptive and Questionable Marketing Practices. Testimony before the Committee on Health, Education, Labor, and Pensions, US Senate. GAO-10-948T. *US Government Accountability Office.*

Malouff, J.M., Thorsteinsson, Schuttle, N.S., Bhullar, N. and Rooke, S.E., (2010). The five-factor model of personality and relationship satisfaction of intimate partners: A meta-analysis. Journal of Research in Personality, 44(1) (2010), pp. 124-127.

Maya-Jariego, I., (2021). Building a structural typology of personal networks: Individual differences in the cohesion of interpersonal environment. Social Networks, Vol 64, pp. 173-180, ISSN 0378-8733, https://doi.org/10.1016/j.socnet.2020.09.006.

Maya-Jariego, Holgado, D., (2015). Living in the metropolitan area. Correlation of interurban mobility with the structural cohesion of personal networks and the originative sense of community. Interv. Psicoso., 24(3), pp. 185-190.

McCrae, R. R., Costa, P.T. (2003). Personality in adulthood: A five-factor theory perspective. Guilford Press

Medicare.gov (2022). Medicare. https://www.medicare.gov/what-medicare-covers

Medicaid.gov (2022). Medicaid. https://www.medicaid.gov/medicaid/index.html

Mesoudi, A (2021), Cultural selection and biased transformation: two dynamics of cultural evolution. *Phil. Trans. R. Soc. B*3762020005320200053

Mount Vernon (n.d.). Building the constitution. Retrieved from: https://www.mountvernon.org/george-washington/constitutional-convention/?gclid=EAIaIQobChMIrN2ljeut-AIVvGpvBB216gzkEAAYASAAEgKbD_D_BwE

Nayyar, S. (2011). *Performance and integration of real time location system in healthcare.* (Order No. MR81383, University of Calgary (Canada)). *ProQuest Dissertations and Theses,* , 178. Retrieved from http://search.proquest.com/docview/912999285?accountid=35812. (912999285).

National Liberty Health Alliance (n.d.). Meet Your Strawman. Retrieved from: www.nationallibertyhealthalliance.org

National Institute of Justice. (2019). Human Trafficking. Retrieved from: https://www.nij.gov/topics/crime/human-trafficking/pages/welcome.aspx

Phares, W. (2014). *Future jihad: Terrorist strategies against America.* St. Martin's Press.

Promades, F. C. (2012). *Influencing graduation rates through resource allocation: A correlation analysis of institutional expenditures and six-year graduation rates at private colleges and universities in New England.* (Order No. 3507787, Johnson & Wales University). *ProQuest Dissertations and Theses,* , 137. Retrieved from http://search.proquest.com/docview/1016089815?accountid=35812. (1016089815).

Robert G. L, Douglas J. B, Steven J. F, (1999). Understanding the dynamics of leadership: The Role of Follower Self-Concepts in the Leader/Follower Relationship, Organizational Behavior and Human Decision Processes, 78 (3), Pgs. 167-203, ISSN 0749-5978, https://doi.org/10.1006/obhd.1999.2832

Roberts. S, Wilson. R., Fedurek, P., Dunbar, R.I., (2008). Individual differences and personal social network size and structure: Personality and Individual Differences. 44 (4), pp. 954-964. ISSN 0191-8869, https://doi.org/10.1016/j.paid.2007.10.033.

Rosenfeld, R. (2012). Culture, people, and innovation. *Research Technology Management, 55*(2), 13. doi:10.5437/08956308X5502007

Rozgonjuk, D., Sindermann, C., Elhai, J.D., and Montag, C., (2021) Individual differences in Fear of Missing Out (FoMO): Age, gender, and the Big Five personality trait domains, facets, and items, Personality, and Individual Differences, V (171) 110546, ISSN 0191-8869, https://doi.org/10.1016/j.paid.2020.110546.

Sanders, A. E., (2013). A gap in the affordable care act: Will tax credits be available for insurance purchased through federal exchanges? *Vanderbilt Law Review, 66*(4), 1259-1307. Retrieved from http://search.proquest.com/docview/1372734472?accountid

Sanger, M. (2008). Getting to the roots of change: performance management and organizational culture. *Public Performance & Management Review, 31*(4), 621-653.

Santana, M. T. (2018). Trafficked in Texas: Combatting the Sex-Trafficking Epidemic Through Prostitution Law and Sentencing Reform in the Lone Star State. *Vanderbilt Law Review, 71*(5), 1739–1773. Retrieved from http://search.ebscohost.com.contentproxy.phoenix.edu/login.aspx?direct=true&db=lgs&AN=132702858&site=eds-live

Shaffer, R. (2022). US Homegrown Political Violence and Terrorism. *Terrorism and Political Violence* Vol. 34 (1). https://doi.org/10.1080/09546553.2021.2017176

Shy, O. (2022). A Model of Student Loans and the Burden of Tuition. *Available at SSRN 4141642.*

Spender, J-C. (1998). The dynamics of individual and organizational knowledge. In C. Eden and J-C Spender (Eds). Managerial and organizational cognition. Theory, methods, and research (pp. 13-39). Sage Publications, Inc.

Tomes, L. (2021). 6 key causes of the American revolution. Retrieved from: https://www.historyhit.com/causes-of-american-revolution/

Wang, Z. (2022). Are There Compelling Reasons to Consider a Central Bank Digital Currency for the US?. *Richmond Fed Economic Brief, 22*(20).

Wendt, L. P., Wright, A. G. C., Pilkonis, P. A., Woods, W. C., Denissen, J. J. A., Kühnel, A., & Zimmermann, J. (2020). Indicators of Affect Dynamics: Structure, Reliability, and Personality Correlates. *European Journal of Personality, 34*(6), 1060–1072. https://doi.org/10.1002/per.2277

White, C. (2012). A Comparison of Two Approaches to Increasing Access to Care: Expanding Coverage versus Increasing Physician Fees. *Health Services Research, 47*(3pt1), 963-983. doi:10.1111/j.1475-6773.2011.01378.x

White House.gov. (n.d). Presidents. Retrieved from: https://www.whitehouse.gov/about-the-white-house/presidents/barack-obama/

Widiger, T.A. (2011). Personality and psychopathology. World Psychiatry, 10(2), pp. 103-06, 10.1002/j.2051-5545. 2011.tb00024.x

World Population Review (2022). Continents in North America. Retrieved from: https://worldpopulationreview.com/country-rankings/countries-in-north-america

Yar, H. A. (2022). Spotlighting a Terrorist: A Review of Racial Profiling in Terrorist by John Updike. *Pakistan Journal of Humanities and Social Sciences, 10*(2), 571-579.

Zhou, W-X., Sornette, D., Hill, R. I.M., Dunbar M., (2005). Discrete hierarchical organization of social group sizes: Proceedings of the royal society of London, 272B, pp. 439-444

Chapter 12

The Leadership Question of the 21st Century

A Futuristic View of Africa

By Anthony Obi Ogbo

INTRODUCTION

This chapter summarizes the fundamental concerns faced by Africa in the 21st century and explores the potential creation of strategies for governance transformation. The structure of governance should facilitate the effective delivery of public service, through a systematic approach that engages with various phases of civil society, community action, social capital, charity, and efforts to combat corruption. These concepts reflect the World Bank's six dimensions of governance: voice and accountability, political stability and absence of violence/terrorism, government effectiveness, regulatory quality, rule of law, and control of corruption (Brown et al., 2018).

There are numerous reasons why Africa as a continent might be considered synonymous with economic, social, and political profligacy. The continent is home to a total of 54 independent countries, three dependencies, and one disputed territory (see Table 1). The United Nations divides Africa into five subregions for statistical purposes: these are Northern Africa, Eastern Africa, Western Africa, Middle Africa, and Southern Africa (worldatlas.com, 2021).

OVERVIEW

Africa's abundance of natural and human resources makes it an area of focus for global investment ventures. The distinct and admirable qualities and opportunities offered by this continent's people and resources are unquestionably both multifold and lucrative. However, the challenge of bad governance

Table 12.1 The African sub-regions and their countries

Sub-region	*Countries*
Northern Africa	Algeria, Egypt, Libya, Morocco, Sudan, Tunisia
Eastern Africa	Burundi, Comoros, Djibouti, Eritrea Ethiopia, Kenya, Madagascar, Mozambique, Malawi, Mauritius, Rwanda, Seychelles, Somalia, South Sudan, Tanzania, Uganda, Zambia, Zimbabwe
Western Africa	Benin Burkina Faso, Cabo Verde, Gambia, Ghana, Guinea, Guinea Bissau, Ivory Coast, Liberia, Mali, Mauritania, Niger, Nigeria, Senegal, Sierra Leone, Togo
Middle Africa	Angola, Cameroon, Central African Republic, Chad, Congo, DR Congo, Equatorial Guinea, Gabon, Sao Tome, and Principe
Southern Africa	Botswana, Eswatini, Lesotho, Namibia, South Africa

Source: https://www.worldatlas.com

remains a stumbling block. The 2021 Corruption Perceptions Index confirms the persistence of deleterious fraud levels amid human rights abuses and democratic uncertainties. A lack of ethical leadership remains widespread, resulting in a corrupt governance culture that pervades the system as a whole.

In light of the above, it may be worthwhile to examine not only the culture and structure of corruption in Africa, but also the persistence of critical incidents and patterns of barbaric insurrection against the ruling class. Researchers argue that inequality among ethnic groups can trigger wars and civil strife. In the continent, disparities in the allocation of resources or power among groups contribute to the unanticipated interruptions to democratic power borne out of multiple incidents and a series of military coups. For decades, most African countries have experienced unprecedented constitutional crises and destructive dictatorships. This trend can be attributed to the inability of the continent's leaders to evenly manage or distribute resources. Thus, the abuse of leadership values that is rampant in Africa is indubitably unconnected with tribalism, sectionalism, schismatic decision-making cultures, and systemic inequality among ethnic groups and/or geopolitical zones.

It is evident that the structures governing most African regions are not consistent with the ideals of the institution of true democracy. Indeed, the continent has suffered due to a lack of good leadership, resulting in accumulated challenges ranging from the devastation wrought by climate change to unforeseen disease outbreaks. Remarkably, it took the COVID-19 pandemic to significantly expose Africa's leadership woes and its leaders' complete unpreparedness for tackling dire unforeseen challenges.

CULTURE AND STRUCTURE OF CORRUPTION

The dynamics of corruption were explained extensively by Mbaku and Sun (2008), who contended that a nation's level of corruption could be determined with reference to certain specific factors: (a) the balance between economic and political power; (b) the magnitude of autonomy afforded to elites; and (c) the level of communication between these officials and their constituents. Thus, in a democracy, a schismatic approach to resource management can weaken the political structure and create a fertile ground for corruption. The authors further affirmed that corruption is likely to be minimal in places where economic freedom is constitutionally guaranteed.

In Africa, a lack of ethical leadership continues to stymie economic development and progress. Dennis Mark Prager, an American conservative radio talk show host and writer, noted that it is not poverty, not lack of riches, and not racism, but rather corruption, which is Africa's greatest problem (BrainyQuote, 2022). Corruption is a betrayal of public trust. A research paper by Mbaku and Sun (2008) clarified that public trust could be betrayed as a result of the behavior and performance of the duties of a public office, exploitation resulting from the economic concept of exchange, and bribery disguised as action in the public interest. Pollitt (2016) notes that in most developing countries, corruption remains a grave obstacle to achieving effective governance. Over the centuries, world leaders have increasingly recognized corruption as an extensive global challenge. Leaders have battled in vain against poverty, inequality, terrorism, conflict, and other calamities triggered by high-level unscrupulous practices.

However, the major issues with the African system include not only its corrupt governance culture, but also the quality of leaders entrusted with the mandate of constructing and upholding an ethical culture. For example, in Nigeria, the leader of an agency tasked with investigating and prosecuting economic and financial crimes was arrested and charged for transgressions completely contradictory to his official mandate. Local media reported that in July 2020, Ibrahim Magu, the Acting Chairman of Nigeria's Economic and Financial Crimes Commission (EFCC), was arrested and charged for "actions prejudicial to state security" (Premium Times, 2020). In July 2021, a United States court issued a warrant for the arrest of Nigeria's Deputy Commissioner of Police, Abba Kyari, on charges related to wire fraud and money laundering (Washington Post, 2021). In May 2022, EFCC arrested the serving Accountant General of the Federation, Ahmed Idris, for money laundering (EFCC, 2022).

In South Africa, former president Jacob Zuma was arrested after a constitutional court found him guilty of contempt for failing to appear before a commission investigating corruption accusations related to his time in office. He was ultimately sentenced to 15 months in prison, capping the stunning downfall of a once-lauded freedom fighter who had battled the apartheid regime alongside Nelson Mandela (Eligon, 2021).

In Ghana, in May 2020, local media reported that a high court in Accra sentenced three former senior government officials to jail terms for purchasing spyware products from the controversial Israeli surveillance firm NSO Group. According to reports, the court affirmed that these officials had caused significant financial losses to the state due to their $4 million purchase of NSO Group's signature Pegasus spyware (Times of Israel, 2020).

In July 2021, France's highest appellate court upheld a guilty verdict against the son of Equatorial Guinea's president for embezzlement, paving the way for the potential return of tens of millions of dollars to the country's people. Teodoro Obiang Mangue, also the vice president of the Gulf of Guinea nation at the time, was handed a three-year suspended sentence and a 30 million euro (USD $33 million) fine at the end of his trial in absentia in 2020. Luxury assets seized in France during the investigation were ordered to be confiscated (France24, 2021). Perhaps most troubling, in light of the seemingly endless nature of these offenses, is that most of those engaged in such corrupt practices are leaders entrusted with managing the people and their resources.

Corruption Perceptions Index

The threat of moral laxity in the leadership sector in Africa shows no signs of abating. Figures from the 2021 Corruption Perceptions Index (CPI) confirm persistently high levels of fraud amid human rights abuses and democratic uncertainties. Regretfully, 90 percent of countries in Africa score below 50 on this index (Transparency International, 2022). The CPI coordinates the global corruption ranking and measures how corrupt each country's public sector is perceived to be, according to experts and businesspeople (Transparency International, 2022).

According to figures and analysis of the 2021 CPI by Transparency International, Sub-Saharan Africa demonstrated no significant improvement, with an average score of 33 out of 100. Worse yet, 44 out of 49 countries assessed on the index still scored below 50.

Corruption in the executive sector can be mitigated. The major problem with most African regions is that nations exist within systems that are vulnerable to amoral practices. Fagbadebo and Dorasamy (2021) correctly stated that the nature of political leadership in Africa rewards a proclivity for the abuse of power. These authors argue that the political elite promotes corruption and employs abuse of power through the instruments of governance by suppressing statutory institutional oversight structures. In their final proposition, Fagbadebo and Dorasamy (2021) contend that leadership or governance crises induced by corruption and mismanagement of public resources will escalate continuously if citizens continue to support unethical practices on the part of their political leaders.

Table 12.2 Corruption Perceptions Index: Africa

	Countries	Scores /100	Ranks /180		Countries	Scores /100	Ranks /180
Ten best African countries					Ten worst African countries		
1	SEYCHELLES	70	23	1	SOUTH SUDAN	11	180
2	BOTSWANA	55	45	2	SOMALIA	13	178
3	RWANDA	53	42	3	EQUATORIAL GUINEA	17	172
4	NAMIBIA	49	58	4	DRC (CONGO)	19	169
5	SAO TOME/ PRINCIPE	19	169	5	BURUNDI	19	169
6	SOUTH SUDAN	11	180	6	CHAD	20	164
7	SOMALIA	13	178	7	SUDAN	20	164
8	EQUATORIAL GUINEA	17	172	8	GUINEA BISSAU	21	162
9	DRC (CONGO)	19	169	9	CONGO	21	162
10	BURUNDI	19	169	10	ERITREA	22	161

Source: Transparency International (2022)

Coups and Crises

Another major issue posing an unprecedented challenge throughout the continent is the inability of leaders to fairly manage or distribute resources. The argument presented by Cederman et al. (2011) that inequality among ethnic groups could trigger war or civil strife finds purchase in the realms of leadership in Africa. Houle (2015) and Houle and Bodea (2017) provide a similar account, stating that disparities in the allocation of resources or power among groups could trigger the unanticipated transition from democracy to autocracy. There is little wonder, then, that serious incidents and coups persist in Africa. This might explain why, for decades, most African countries have fluctuated between shaky democracies and destructive dictatorships. On January 31, 2022, for example, the New York Times reported on how military leaders had toppled the governments of Mali, Chad, Guinea, Sudan, and Burkina Faso over the previous 18 months.

The history of coups in Africa in the past and present is a clear indication that all is not well with leadership in the continent. The proliferation of coups has become a governmental culture often marked by civil unrest and wars, a trend that remains consistent to date. As noted above, over the past 18 months or so, Africa has witnessed seven coups or coups attempts. Such occurrence and events portend grave danger for Africa's nascent democracy. High-ranking military officers in Burkina Faso, Chad, Guinea, Mali, and Sudan have successfully overthrown existing legitimate democracies and assumed power; in Niger and Guinea-Bissau, similar attempts were made, although these were unsuccessful. United Nations Secretary-General Antonio Guterres called it "an epidemic of coup d'états (Maclean, 2022)

Table 12.3 Sub-Saharan Africa Highlights

The Sub-Saharan Africa average CPI score is 33, the lowest in the world, and 44 African countries score less than 50:

- Seychelles (70) tops the region, with Cabo Verde (58) and Botswana (55) the distant runners-up.
- Equatorial Guinea (17), Somalia (13), and South Sudan (11) score lowest in the region.
- Botswana (55), Lesotho (38), Eswatini (32), Niger (31), Nigeria (24), Comoros (20), and South Sudan (11) are all at historic lows this year.

Over the last decade, 43 countries in the region have either declined or made no significant progress:

- Since 2012, Botswana (55), Liberia (29), Mali (29), and South Sudan (11) have seen significant declines in their CPI rankings.
- Only six countries in the region have significantly improved their scores over that period: Seychelles (70), Senegal (43), Ethiopia (39), Tanzania (39), Côte d'Ivoire (36), and Angola (29).

Source: Transparency International (2022)
Note: The CPI ranks 180 countries and territories by their perceived levels of public sector corruption on a scale of zero (highly corrupt) to 100 (very clean)

This latest wave of coups began with Mali in 2020, when a military regiment took over the country's government. The coup leaders took advantage of the public outcry over a controversial parliamentary election. Citing the government's failure to curtail violence and extremism, they arrested the President, Ibrahim Boubacar Keita, and forced him to resign on state television. Second and third coups in Mali followed nine and ten months later, respectively. The second coup took place in May 2021, when the Malian Army led by Vice President Assimi Goïta removed the regime of President Bah N'daw and took over the country's affairs.

Meanwhile, Mali's first coup was in 2012: some rebel Malian soldiers, dissatisfied with the government's handling of a Tuareg insurgency in the north of the country, removed sitting president Amadou Toumani Touré and resumed power. A thorough assessment of Africa's pattern of governmental overthrow reveals that soldiers involved in military takeovers tend to share certain commonalities. Very often, they highlight corruption, insecurity, political and economic instability, and fragile democratic sectors as reasons for insurrection.

A substantive argument focusing on how to discourage a rebellious military governmental takeover was expounded in a study by Albrecht and Eibl (2018), titled "How to Keep Officers in the Barracks". The researchers argue that elite- and combat-officer coups involve fundamentally different execution processes and, in turn, require distinctive coup-proofing strategies. Coups led by elite officers are largely driven by the political ambitions of these high-level officers and the internal dynamics of authoritarian regimes. Combat officers, however, see themselves as ordinary citizens. This is because combat officers—unlike their elite counterparts, who are generally upper-class—derive meaningful economic benefits from general welfare expenditure. The authors argue that complete political reforms that prioritize system liberalization could significantly diminish the likelihood of combat-officer coups.

The good news regarding the state of leadership patterns in Africa is that democratic governance is gradually replacing Africa's dictatorial regimes. However, Afolayan (2011) noted that democracy, as understood on the continent, seems in no way better than dictatorship if ethical rectitude, social responsibility, and the fair distribution of a country's wealth are taken as yardsticks. Proponents of Africa's dictatorships are often the stakeholders of that system and would argue that real democracy and credible elections are too messy, too unpredictable, and too expensive, with uncertain outcomes (Daily Nation, 2017).

Strategists wishing to quash a growing insurrection in an African democracy might begin by evaluating the structure of power and governance. The inability of Africa's leadership to properly sustain the ideals of democracy often creates rugged terrain that exacerbates system uncertainties and provokes chaos. Ford and Versi (2022) contend that democracy in most of the African countries where coups have taken place has been relatively weak. Thus, observable traits such as limited tolerance of political opposition, restrictions on freedom of expression and independent media, and a judiciary heavily influenced by the government are major indicators that rebellion may occur. Furthermore, coup leaders often promise a short-term correctional regime, but in most cases end up retaining a prolonged hold on power (Ford & Versi, 2022). However, the Daily Nation opined in 2017 that it is better to have a good leader for a short period than a bad one for a long period.

Climate Change

The present state of the climate in Africa has been attributed to the continent's poor leadership standards. According to the World Meteorological Organization (2020), increasing temperatures and sea levels and changing rainfall patterns, coupled with extreme weather, have not only threatened livelihoods but also posed additional challenges to all aspects of socio-economic growth.

Accounts from the State of the Climate in Africa 2019 Report, a multi-agency publication produced by the World Meteorological Organization (WMO), provide a reflection of current and future environmental trends, along with analysis on how they might affect the economy and other relevant sectors of development. Increasing temperatures and sea levels, changing patterns of rainfall, and extreme weather have endangered livelihoods, posing challenges to all aspects of socio-economic growth. UN Climate Change News (2020) reported that 2019 was among the three warmest years on record for the continent and that this trend was expected to continue.

There was also a concerning revelation regarding significant regional variability in sea-level trends around Africa. For example, Tropical Cyclone Idai was among the most destructive tropical cyclones ever recorded both on the continent and in the southern hemisphere, causing catastrophic damage and a humanitarian crisis in Mozambique, Zimbabwe, and Malawi; high casualties and the displacement of hundreds of thousands resulted, leaving more than 1,300 people dead and many more missing (UN Climate Change News, 2020).

There have also been major issues related to food security triggered by the devastation of climate change. The number of undernourished people in drought-prone sub-Saharan African countries increased by 45.6% since 2012 following serious damage to agricultural development, which is the backbone of Africa's economy and accounts for the majority of livelihoods across the continent. The health impacts of fluctuating but increasing temperature and changes in rainfall patterns also place the African population at risk. Warmer temperatures and higher rainfall increase habitat suitability for biting insects and the transmission of vector-borne diseases such as dengue fever, malaria, and yellow fever. The dire economic impact, according to the International Monetary Fund, is particularly noticeable in regions with relatively hot climates, where a disproportionately large number of low-income countries are located (World Meteorological Organization, 2020).

Africa is both the continent most vulnerable to climate change, heating up faster than the global average temperature, and the least able to afford the cost of mitigating policies and practices. One potential solution was articulated in a recent editorial published in the Ugandan Daily Monitor (2022). An excerpt reads, in part:

> "In Africa, we have the technical know-how and the indigenous knowledge to reproduce natural resources. We have the ambition and the expertise to lead a globally coordinated effort. And we have the biodiversity to re-establish thriving habitats. What we do not yet have is a consensus among African nations that we must do this for ourselves and that the wealthy nations must help finance our efforts because they will benefit, too."

Resolutions made at the 2021 United Nations Climate Change Conference (COP26), held at the SEC Centre in Glasgow, Scotland, provided another opportunity for African leaders to mitigate the wrath of climate change. The conference, in which almost 200 countries were represented, explored comprehensive strategy options that would accelerate climate action in this decade, and finally completed the Paris Rulebook (The UN Climate Change Conference, 2021). The Paris Agreement enshrines a global commitment to take multilateral action against climate change. The Glasgow resolution addresses mitigation (reducing emissions), adaptation (helping those already impacted by climate change), finance (enabling countries to deliver on their climate goals), and collaboration (working together to deliver even greater action) (COP 26, 2021).

In another paper, entitled "Addressing Africa's dual challenges: Climate change and electricity access," Lioko (2022) contended that a global response to climate change must incorporate the context of historic emissions, citing China, Europe, and the United States as regions that bear the most responsibility for greenhouse gas emissions. In other words, prioritizing the transition to renewables and imposing higher emissions reduction requirements on these advanced countries will ease the burden on those nations (especially those in Africa) that still need to make use of a variety of power generation methods to increase energy access (Lioko, 2022).

Wrath of the Pandemic

The COVID-19 pandemic, which emerged in 2020 and has subsequently ravaged the globe for more than two years, further exposed Africa's leadership woes and the lack of preparedness among its leaders to tackle dire unforeseen challenges. Even after the first case was confirmed in Egypt on February 14, 2020, the continent remained slow to act, paving the way for calamitous consequences. For example, as of November 22, 2021, Africa had already recorded 222,276 deaths, with South Africa facing the highest number of casualties (89,584 deaths), representing roughly 40.3 percent of the total (Statista.com, 2022). Figures also show that Tunisia followed South Africa, with 25,347 victims (almost 11.4 percent of the overall deaths in Africa), while Egypt accounted for around nine percent of the continent's casualties with 19,991 victims. As of May 16, 2022, the World Health Organization reported 8,143,104 cumulative cases on the continent, broken down to 1,857 new confirmed cases, five new deaths, 169,748 cumulative deaths, 7,510,041 recoveries, and 463,315 active cases (COVID-19, WHO African region, 2022).

A research article by Mbopi-Keou et al. (2020) ties Africa's inability to fully manage the onslaught of COVID-19 to two key factors: "the structural inadequacy of health systems and the serious deficiencies in the African approach to the response" (p. 2). Exacerbating the rampage of COVID-19 across the globe was the continent's inadequate response to its high-incidence nations. A detailed report by Odeyemi et al. (2020) highlighted extremely low testing figures and ineffectual laboratory measures as evidence of sub-optimal success in the continent's fight against COVID-19.

African leaders have the potential to prepare the continent against unforeseen disasters. In great civilizations, significant challenges such as pandemics could be transformed into undeniable opportunities to rise to the challenge; in the same way, as Francois-Xavier et al. (2020) noted, African scientists could equally rise to meet the challenges of unforeseen health related epidemics.

The lessons of the pandemic have revealed the necessity of integrating digital technology into the health sector to mitigate unforeseen challenges at a fundamental level. During critical periods of the COVID-19 pandemic, most African regions struggled to coordinate accurate information updates. The success of communication management in every epidemic depends on the proper dissemination of information between health agencies and the community, as misinformation may lead to a disorganized response process and increased fatalities (Adesanya, 2020).

In the era of mobile technology innovation, Odeyemi et al. (2020) recommend that greater and improved healthcare financing, along with the adoption of mobile testing laboratories, especially in rural communities, could inspire proper epidemic preparedness. Indeed, in Africa, technology could serve as a powerful medical tool to support the health sector at all levels of care. According to Sharma et al. (2022), mobile phones have proven to be a useful communication tool not only for data transmission but also for effective healthcare delivery. This was evident during the COVID-19 pandemic, when mobile technology played an important and instructive role in curtailing the spread of the virus.

DISCUSSIONS: A VIEW OF THE FUTURE

A long thread of issues and analyses about Africa boils down to one concept: "leadership." Based on the challenges discussed above, the explosion of coups in the continent in recent years and prior centuries is a clear indication that all is not well with African leadership; the present state of the climate has been attributed to the continent's poor leadership standards; furthermore, the COVID-19 pandemic vividly exposed Africa's leadership woes and the lack of preparedness among its leaders to tackle dire unforeseen challenges. Therefore, a thorough reflection on the discipline of leaders, the process of leading, and the values of effective leadership has become crucial to developing a strategic solution that will usher in a transformational approach to Africa's governance efficacy.

Authors, scholars, and researchers alike have blamed the state of Africa's economy on leadership inadequacy. This challenge cuts across all African countries, leaving the continent's politics, people, and resources in shambles. Critical issues regarding the lack of accountability, corruption, marginalization, disregard for the rule of law, tribal and sectarian unrest, and incompetence in the leadership sector have been addressed at length in numerous journals. For example, former United Nations secretary-general Kofi Annan blamed African leaders for the continent's leadership devastation during his tenure in 2005 (AllAfrica, 2016). Addressing the then-Organization of African Unity (now African Union), Annan accused the leaders of not doing enough to address investment and development challenges. According to Annan, "We have mismanaged our affairs for decades, and we are suffering the accumulated effects" (AllAfrica, 2016).

In what seems to be a collective pattern, most African countries have suffered under a lack of social, economic, and political development, born from a lack of ineffective political leadership. At some point, constituents under democracies and those under authoritarian rule alike must contend with ineffectual execution of official responsibilities, as leaders fumble with their political power and watch their constituents slide into a near-irreparable economic slump. According to Afolayan (2011), "Both modes of governance (democratic and military) [lead to] nuclear meltdowns and backwardness, despite the heroic fortitude, optimism, and ingenuity of [the continent's] hardworking populace" (p. 38).

The Discipline of Leaders, Leading, and Leadership

In the context of this paper, leadership can be associated with guidance, control, and facilitation. While early studies of leadership focused on the traits of so-called great leaders, contemporary approaches tend to focus on the process of leadership, including behavioral and stylistic approaches, situational and contingency theories, transformational and transactional leadership, leader–member exchange theory, post-heroic approaches, and complexity leadership theory, to name the key approaches (Jeanes, n.d). Leadership can also be conceived of as a role in a group assigned on the basis of personal characteristics, experience, or tradition; in other words, a leader can play a major part in making group decisions, motivating participants, and coordinating group actions (Kent, 2006).

Hence, to meet the leadership challenges that are currently impeding the progress of development in most African countries, a suitable leader could be a prudent risk-taker who directs tough innovative, and transformative solutions involving unexpected outcomes (Tichy & Devanna, 1990). Such a leader must inspire change, motivate constituents, and carry them along toward the finish line (Gallos, 2006, 2007; Jones, 2010; Yukl, 2013; Nahavandi, 2012) of operational success.

This chapter explores a view of the present and future in light of the vacuum created by the low standards of leadership in homeland Africa. Indeed, one of the keys to an organization's long-term success lies in the ability of the leadership to rise to tough challenges. Thus, African leaders must effectively develop a clear democratic vision, construct a supporting organizational structure immune from systemic dichotomy and prejudice, and commit to a strategic plan (Yukl, 2013; Nahavandi, 2012) to squarely address the economic and sociopolitical interests of the people.

To explore Africa's leadership question in the 21st century—or perhaps, to analyze the quality of current leaders in the continent regarding specific matters of governance—the foundation and theories of leadership must be put in the right context. Studies of early leadership theories attributed effective leadership to personality traits. However, an evidence-based reflection by Wren (1995) ascribed leadership to the strategic interaction of three indispensable elements: the leaders, the followers, and the surrounding situation. Wren's thoughts have also been echoed by great leaders over the years. For example, "General George Patton's avowal that one can either lead, follow, or get out of the way dovetails the classification of leadership theory and research through three relevant variables: leaders, followers, and the situation" (Ogbo, 2015; Ogbo, 2016, p. 155). However, the conceptual issues related to politics must be put in context if they are to have a major influence on decision-making effectiveness.

Theories of leadership notwithstanding, politics conventionally defines the capacity or power of leadership, the players involved, how leadership is exercised, and how it is socially constrained (Walter, (2007). Furthermore, in democratic governance, leadership can be influenced not only by the art and science of government, but also by the constitutional structures that delineate the leader's decision-making prowess. For instance, the application of politics in leadership sets the tone for the distribution of resources and determines how difficulties or incongruities can be best resolved (Allison, 2018). It is no wonder that Afolayan (2011) observed that democracy could thrive if ethical rectitude, social responsibility, and the fair distribution of resources are taken as yardsticks.

FINAL REFLECTIONS

True leadership entails respect for people, process, and the rule of law. Leadership in Africa may remain a troubled institution until nations restore thorough respect for the rule of law. Dictatorship in Africa is not only a tactic for the military juntas; its authoritarian tenets are invoked when leaders invent constitutional clauses to overstay their tenures, or possibly subvert the process to ensure they remain in office. In short, this is undemocratic. Most African democracies are indeed victims of leaders who would not leave office. For instance, President Yoweri Museveni of Uganda was reelected on May 12, becoming one of the longest-serving African leaders (Africa's longest serving Presidents, 2021). The Ugandan Daily Monitor (2021) wrote: "His reelection last Thursday will take Mr. Museveni, who has ruled Uganda since January 1986, to 40 years in power by the next General Elections in 2026. The President, now aged 76, will be 81 by then"

But Uganda is not the only country struggling under a leader who refuses to simply serve and leave. To mention only a few examples, Teodoro Obiang has been leader of Equatorial Guinea since 1979; 87-year-old Paul Biya of Cameroon has been serving since 1982; Denis Sassou Nguesso of the Republic of Congo served from 1979–1992, then returned in 1997 and remains in power today (Reuters.com, 2021; Nairobi News, 2021).

The major question remains: how can Africa prosper under structures that undermine core organizational transformation standards, when leaders subvert the laws to feed their lust for power and greed? How can democracy thrive under such tyrannical and inequitable extension of tenure? How can African leaders deliver effective public service in a way that accords with people's hopes and aspirations? How could these leaders position their governance system to meet the challenges of the trending global crisis?

This chapter focuses on effective leadership standards as it explores the present challenges and potential future developments of leadership in Africa. A one-size-fits-all solution is not offered; rather, these passages have discussed relevant concepts with a view to creating flexible approaches to mitigation. Such flexibility entails the application of strategies based on the specific challenges or economic and socio-political environments involved. As Techo (2021) noted, "by being flexible, the leader can find different approaches to solve the same problem—thus, the leader acts in a transformational manner," (p. 6).

The major issues Africa faces in the 21st century and an overview of their implications for the future have been explored above in the context of creating avenues for effective governance. From the explosion of coups in the continent in recent and less recent history, the climate change-related challenges that can be attributed to the continent's poor leadership standards, and the pandemic that emerged in 2020 and has ravaged the globe for over two years, there is no doubt that Africa's leadership afflictions are connected with the continent's inability to manage dire unforeseen challenges.

To meet such challenges, leaders must be equipped with transformational individual qualities, including personality, values, confidence, expertise, influencing tactics, and related relevant behaviors (Yukl, 2013). Paradigmatic changes come with challenges. Dealing with the very difficult challenges of governance requires strategies suitable for achieving cultural transformation (Scott & Davis, 20011). Unfortunately, in most African countries, the principles needed to promote change are almost inexistent. However, given the trends in global transformation, it is time to move this continent from its present state into the projected future and to inspire growth (Jones, 2010). Moreover, any movement towards change will only be effective when African leaders understand the significant associated complications and other barriers that must be mitigated.

EPILOGUE: EPISTEMOLOGY OF LEADING

Due to long-standing and egregious abuses of power, the sense and discipline of leadership and the act of leadership in Africa are currently hampered by severe structural inadequacies. Authors, researchers, and theorists of leadership have for many years crafted their definitions from countless perspectives, justifying the conceptual relevance of each definition with substantial empirical evidence. In fact, most definitions are harmonized with reference to theoretical models of organizing people to manage their environment, politics, and resources. However, the proper identification and deconstruction of inadequacies in practice have gradually been left out of consideration. For example, most definitions of leadership might end up exploring the epistemic standards associated with leading or managing success. Perhaps, to an extent, this paper considers the deeply senseless actions taken by African leaders to the detriment of their constituents. As a solution to Africa's leadership catastrophe, however, this paper invokes the definition, challenges, and practice of leadership from a holistic perspective, significant theories, and concepts to convey the meaning and application of leadership in practice.

The purpose of this paper is to explore the leadership question in the 21st century, adopting a forward-looking perspective and creating avenues for improving governance system effectiveness. To open up truly constructive discussions that may lead to this mission being accomplished, this paper provides remedies for the crisis of leadership that extend beyond the conceptual boundaries of organizing people, their hopes, and their aspirations. Theoretical principles explain the process of leading or managing and offer substantial clues as to how African leaders could harness their resources and translate their communal crises into a rich array of new economic possibilities. In a strategic reversal, rather than asking and explaining what leadership is or should be, the approach adopted herein is to ask and convey what leadership *is not* and *should not* be, and hence what African leaders might be doing to avoid creating an innovative leadership culture.

Hence, it might be pertinent to proceed with an analysis of what leadership is not by first asserting that leadership is not voodoo. It is not a political chess game, or a matter of fanfare; it is a social science. In fact, in the broadest sense, the application of leadership must *make sense*. The process must facilitate philosophical clarification in meaning and practice. It must reflect the interests of both leaders and followers. As noted by Malakyan (2014), "Human beings are born with abilities to lead and follow (organic leadership and followership). Leaders and followers share a common humanity with their unique personalities and characteristics" (page 17). Therefore, it could be argued that the tradition of leadership in Africa must not only encompass a collaboration between the leader and the follower but must also appeal to an expansive consideration of communal interests. Nonetheless, if the exercise of leading people does not manifest as a calm and commonsense approach to understanding and addressing the challenges faced by the masses or carrying them along as partners in the decision-making business, it is not leadership.

In most African countries, the marginalization of tribes and regions has incited various rebellions or revolts against the ruling class. But there is also good news. Malakyan (2014) invoked the leader–follower trade (LFT) approach to convey how a non-static and organic approach could influence different situations or organizational settings towards mutual respect, empowerment, and effectiveness. Thus, if the governance practice does not encourage an environment that fosters wisdom —where leaders learn from followers and followers learn from leaders; where leaders uplift followers and followers uplift leaders; or, perhaps, where both groups walk side-by-side in solidarity over their shared need to transcend challenges—it is not leadership.

Most African countries are, obviously, in the midst of a dire sociopolitical and economic crisis. In most cases, the masses reach the end of their endurance while waiting eternally for their leaders to deliver the promised reforms to the system. At present, aspiring leaders are motivated solely by a desire to enrich themselves. In most democracies, constitutionally structured power management and transition become an impossible mission, leaving military coups as the only option (Ford & Versi, 2022). However, the executive, judicial, and legislative sectors of leadership have the statutory prerogative to create a constructive governance structure that not only protects the interests of the masses but also opens up a path to uplifting all geopolitical regions. Therefore, if the system is not structured so as to guarantee equitable allotment of resources; if the structure does not impartially protect the political, economic, or social interests of all demographic or tribal subdivisions; if, indeed, the structure does not irrevocably protect the independence of the executive, judicial, and legislative arms in such a way as to curtail decision-making excesses, it is not leadership.

Leadership entails the courage to make bold steps toward establishing long-lasting system structures. Most African countries, in one way or another, are challenged by the contentious politics of governance restructure. Fear of embracing something entirely new, coupled with the struggle for power and resources, complicates the creation of effective leadership and governance structures. In Nigeria, for instance, the demand for governance restructure has dominated the political scene: "From the Aburi Accord in Ghana in January 1967 (Nwokedi, 2022) to the General Abacha constitutional conference of 1995 to the President Obasanjo's National Conference of 2005, and the President Jonathan's Confab of 2014, the reality of agreeing on an all-inclusive basis of nationhood remains undebatable" (All Africa, 2022). Over the years, however, the people have regrettably witnessed the obstructive procrastination that has prevented the rebuilding of the country to reflect the people's hopes and aspirations for one united nation. Therefore, in the quest for structural transformation, if governance practice does not entail the taking of bold legislative risks and facilitating innovative solutions to system restructure and transformation, it is not leadership.

Effective communication is inevitably a key component of political leadership effectiveness. Through open communication, leaders can speed up innovation by expediting, planning, goal-setting, exchanging information, managing risks, and suppressing misinformation and misunderstandings. However, the coordination of every outbreak response strategy in most African countries has been derailed by a lack of proper information dissemination between the health agencies and the community. For example, in his research article titled, "COVID-19 Outbreak in Africa: Lessons and Insights from the West African Ebola Virus Disease (EVD) Epidemics," Adesanya (2020) noted that during the 2014 Ebola outbreak in Nigeria, misinformation regarding the efficacy of saltwater and bathing in it, prevented the virus led to the avoidable hospitalization and death of about five people during the outbreak. The author further noted that "there was widely spread misinformation on how safe it was to accept survivors back into society, resulting in the stigmatization of survivors and the development of post-traumatic stress disorder" (p. 98).

Consequently, Blake et al. (2015) opined that African countries—especially those combating insurgencies, such as the Federal Republic of Nigeria, Mali, Morocco, Democratic Republic of Congo, Uganda, Rwanda, Central African Republic, South Sudan, and Kenya—could invest in the use of Information and Communication Technologies (ICT) for promoting and sustaining national integration. Therefore, if the practice in question does not entail innovation and collaboration among stakeholders of governance, through effective communication, or if the practice does not create a culture of transparency and information control to close the gap between leaders and the masses, it is not leadership.

Leadership is not a one-man process, especially in politics where multiple interests are at play. Political leaders do have constituents, who could conceivably act as their partners in system management. Isolating any geopolitical sector in the decision-making process often results in the formation of a monocratic or tyrannical structure. In Africa, the challenges of dichotomy and marginalization continue to generate political instability and conflict. According to Achot and Bill (2022), in general terms, the main motivations that drive violent regionalist or secessionist movements (which, in most cases, take up arms against a government) are a combination of political and economic factors, aggravated by the marginalization of state policies. Therefore, if the leadership does not consider it an official responsibility to recognize all demographic sectors or geopolitical precincts and adopt them as allies in managing resources, or if the process does not carry both the leaders and the masses along through the transformation venture, in which the fundamental process of embracing something new requires the construction of sociopolitical awareness and shared collective interests, it is not leadership.

Leaders succeed when they manage and ensure the effectiveness of knowledge-based organizational structures. However, leaders must have the intellectual capacity to lead or facilitate knowledge management processes. Kazak (2021) noted that as part of organizational culture, leaders can expedite knowledge management activities by choosing the most appropriate types of infrastructure and technology. Organizations are more innovative when they respect knowledge development and encourage stakeholders to stay ahead of competitors in a knowledge development context (Shane, 2009). In addition to knowledge capacity, a leader must be equipped with the right governance skills; specifically, he must possess the ability to perform certain tasks satisfactorily. Hashimzade et al. (2017) noted that these skills may involve physical dexterity, mental ability, or both, which can be learned either through formal instruction or through career experience. Because democratic leaders are often elected rather than employed, the chance that a leader may be unqualified cannot be ruled out. Similarly, in the event of a forceful military takeover, a country might be left with a leader who is completely ignorant of both leading and governance. Therefore, if the system is left with an executive class that possesses little or no knowledge about managing people and resources, or if the system is stacked with executive team players who have little or no knowledge about the values of innovation and/or no regard for the laws of the land, it is not leadership.

Furthermore, in the African context, where good leadership is highly valuable, it must be noted that longevity in public service might not be enough to succeed in the governance transformation venture. Instead, a leader must be armed with the intellectual capacity required to manage and motivate a transformation culture. Most African leaders have been accused by their constituencies of having limited leadership qualities. However, leadership must represent a beacon of hope, human endurance, individual courage, and resolute moral fortitude. Therefore, if the process is not knowledge-based; if the leader lacks the charisma required to motivate the masses; and if the leadership does not exhibit a propensity to engage in bold experimentation, take on the potential risk of innovation, and manage failure, it is not leadership.

In Africa, the charismatic elements of the act of leading are often downplayed, especially in countries where a capitalistic quest for power and profitability engulfs human empathy. Leadership must have a human face and a capacity for compassion. CNN Wire (2021), when analyzing the success of former South African president Nelson Mandela, noted that Mandala forgave his former captors, flashed luminous smile, and won them over with his decency. This was not so in Nigeria. In his first visit to the United States after being sworn in as president in 2015, President Muhammadu Buhari shocked his country when he attempted to justify his vindictive and flawed nepotism, explaining that it would amount to injustice to treat those who voted for him and those who did not vote for him equally (Sun, 2021). Understanding human emotions could include empathy and social sensitivity. With a high level of clear communication and compassion for the plight of the masses, African leaders could boost loyalty and build a high level of patriotism. Therefore, if the process does not inspire empathy—that is, if it lacks the compassionate characteristics concomitant with the milk of human kindness, or if it does not resonate with the plight of the common people and share their traditional values and collective aspiration—it is not leadership.

In our current era of social, political, and economic uncertainties, the language of leadership could build hope rather than fear, convey love rather than hate and animosity, inspire peace rather than hatred, and accordingly, build unity rather than division. As Mahātmā Mohandas Gandhi noted, "service without humility is selfishness and egotism." Therefore, if the leader exhibits an attitude of arrogance, tweets fear, and adopts the pitiless disposition of a dictator, it is far from leadership.

Most African countries are in dire need of transformation. The current global challenges necessitate the embrace of something entirely new. Thus, the application of leadership must galvanize a culture of transformation. As a leader, it might not be sufficient to rule only by the written laws. African leaders must work as a team in looking beyond the norms to conceive of innovative strategies for moving the continent from its dilapidated present state to an elevated level of economic possibilities. As a team, African leaders must be able to speak the language of transformation with an accent pertinent to innovation and transformation culture. Therefore, if the process does not speak the language of transformation and does not recognize the values of innovation—in other words if the system undermines the necessity for change and undercuts the technology of organizational reformation and progress—it is not leadership.

REFERENCES

Achot, G. A. R., & Bill, D. B. A. (2022). Strategy for political violence in Africa: The case of "marginalizing state" in Sudan. *Journal of Social Sciences & Humanities Researches*, *7*(2), 114–122.

Adesanya, O. A. (2020). COVID-19 outbreak in Africa: Lessons and insights from the West African Ebola Virus Disease Epidemics. *International Journal of Travel Medicine & Global Health*, *8*(3), 96–99. https://doi.org/10.34172/ijtmgh.2020.17

Afolayan, S. (2011). African literature: A showcase for Africa's leadership problems. *Matatu: Journal for African Culture & Society*, *39*(1), 27–39. https://doi.org/10.1163/9789401200 745_003

Africa's longest serving Presidents. (2021, May 14). Nairobi News, NA. https://link.gale.com/apps/doc/A661776780/GIC?u=uphoenix&sid=ebsco&xid=0f7739a

Albrecht, H., & Eibl, F. (2018). How to keep officers in the barracks: Causes, agents, and types of military coups. *International Studies Quarterly*, *62*(2), 315–328. https://doi.org/10.1093/isq/sqx085

allAfrica.com. (2022, March 25). *Nigeria: The Only Way Out Is to Restructure Nigeria.* https:// link.gale.com/apps/doc/A698229380/GIC?u=uphoenix&sid=ebsco&xid=06c4be8e.

Allison, L. (2018). Politics. In *A Concise Oxford Dictionary of Politics and International Relations*. Retrieved June 26, 2022, from https://www.oxfordreference.com/view/ 10.1093/acref/9780199670840.001.0001/acref-9780199670840-e-1042.

Blake, C., Fayomi, O., & Ayo, C. (2015). The use of information and communication technologies for promoting and sustaining national integration in Africa. *Proceedings of the European Conference on E-Learning*, 37–43.

BrainyQuote.com. (n.d.). *Dennis Prager Quotes*. https://www.brainyquote.com/quotes/ dennis_prager_649301

Brown, G., McLean, I., & McMillan, A. (2018). Governance. In *A Concise Oxford Dictionary of Politics and International Relations*. Retrieved May 2, 2022, from https://www. oxfordreference.com/view/10.1093/acref/9780199670840.001.0001/acref-9780199670840-e-1539.

Cederman, L.-E., Weidmann, N., & Gleditsch, K. S. (2011). Horizontal inequalities and ethnonationalist civil war. *American Political Science Review*, *105*(3), 478–495.

CNN Wire. (2021, July 18). *We need Nelson Mandela's example more than ever today. How the South African leader's radical empathy can help the US today.* Retrieved from https://www.cnn.com/2021/07/18/us/nelson-mandela-birthday-lessons-for-us-politics-today/index.html .

COP 26. (2021). *The Global Climate Pact. Glasgow, U.K.* https://ukcop26.org/wp-content/uploads/2021/11/COP26-Presidency-Outcomes-The-Climate-Pact.pdf

COVID-19 (WHO African region) March 16, 2022. Retrieved from https://who.maps.arcgis.com/apps/dashboards/0c9b3a8b68d0437a8cf28581e9c063a9

Daily Monitor. (2022, February 19). *We need bold African leadership.* https://link.gale.com/ apps/doc/A697577383/GIC?u=uphoenix&sid=ebsco&xid=4cc40ad2.

Daily Nation. (2017, March 17). Kenya (Nairobi, Kenya). Retrieved from https://nation.africa/kenya/blogs-opinion/opinion/dictatorships-cannot-solve-african-leadership-problems-374376

Efcc.gov. (2022). *EFCC Arrests Ahmed Idris, Accountant-General of the Federation for N80billion Fraud. Media & Publicity.* https://www.efcc.gov.ng/news/8008-efcc-arrests-ahmed-idris-accountant-general-of-the-federation-for-n80billion-fraud

Eligon, J. (2021, July 7). *Jacob Zuma, Former South African President, Is Arrested.* New York Times. https://www.nytimes.com/2021/07/07/world/africa/jacob-zuma-arrested-south-africa.html

Fagbadebo, O., & Dorasamy, N. (2021). Political leadership, corruption, and the crisis of governance in Africa: A discourse. *African Renaissance, 18*(1), 27. https://doi.org/10. 31920/2516-5305/2021/18n1

Ford, N., & Versi, A. (2022). What is behind the spate of coups in West Africa? *New African, 613*, 26–31.

France24. (2021, July 28). *French court upholds guilty verdict on fraud charges for Equatorial Guinea's VP.* https://www.france24.com/en/france/20210728-french-court-upholds-guilty-verdict-on-fraud-charges-for-equatorial-guinea-s-vp.

Gallos, J.V. (2006, 2007). *Organization development.* A Francisco, CA; Jossey-Bass.

Hashimzade, N., Myles, G., & Black, J. (2017). Skills. In *A Dictionary of Economics.* Retrieved June 3, 2022, from https://www.oxfordreference.com/view/10.1093/acref/9780198759 430.001.0001/acref-9780198759430-e-2882

Houle, C. (2015). Ethnic inequality and the dismantling of democracy. *World Politics, 67*(3), 469–505.

Houle, C., & Bodea, C. (2017). Ethnic inequality and coups in sub-Saharan Africa. *Journal of Peace Research, 54*(3), 382–396.

Jeanes, E. (n.d.). Leadership. In *A Dictionary of Organizational Behaviour.* Retrieved March 13, 2022, from https://www.oxfordreference.com/view/10.1093/acref/9780191843273.001 .0001/acref-9780191843273-e-169.

Jones, G.R. (2010). *Organizational theory, design, and change* (6th Ed.). Upper Saddle River, NJ: Prentice Hall.

Kazak, E. (2021). A conceptual analysis of the role of knowledge management in knowledge leadership. *Journal of Theoretical Educational Science / Kuramsal Eğitimbilim Dergisi, 14*(3), 352–372. https://doi.org/10.30831/akukeg.856221.

Kent, M. (2006). Leader. In *The Oxford Dictionary of Sports Science & Medicine.* Retrieved March 13, 2022, from https://www.oxfordreference.com/view/10.1093/acref/ 9780198568506.001.0001/acref-9780198568506-e-3918.

Lioko, J. M. (2022). *Addressing Africa's dual challenges: Climate change and electricity access.* Brookings. https://www.brookings.edu/blog/africa-in-focus

/2022/01 /25/addressing-africas-dual-challenges-climate-change-and-electricity-access/

Maclean, R. (2022). *Five African Countries. Six Coups. Why Now?*. The New York Times. https://www.nytimes.com/article/burkina-faso-africa-coup.html.

Malakyan, P.G. (2014), Followership in leadership studies: A case of leader–follower trade approach. *Journal of Leadership Studies*, *7*, 6–22. https://doi.org/10.1002/jls.21306.

Mbaku, J., & Sun, Y. (2008). Corruption. In *The Oxford Encyclopedia of the Modern World*. Retrieved May 21, 2022, from https://www.oxfordreference.com/view/10.1093/acref/9780195176322.001.0001/acref-9780195176322-e-375.

Mbopi-Keou, F.-X., Pondi, J.-E., & Bélec, L. (2020). COVID-19: The urgency of the doable for an African scientific leadership. *The Pan African Medical Journal*, *37*(44). https://doi.org/10.11604/pamj.2020.37.44.24401.

Museveni joins Africa's longest serving leaders. (2021, January 18). *Daily Monitor*, Uganda (Kampala, Uganda) Retrieved from https://www.monitor.co.ug/uganda/special-reports/museveni-joins-africa-s-longest-serving-leaders-3261008

Nahavandi, A. (2012). *The art and science of leadership* (6th ed.). Upper Saddle River, NJ: Pearson Education.

Nwokedi, U., (2022). *A Shred of Fear – A Memoir*. Narrative Landscape Press. Lagos, Nigeria

Odeyemi, F. A., Adekunle, I. A., Ogunbanjo, O. W., Folorunso, J. B., Akinbolaji, T., & Olawoye, I. B. (2020). Gauging the laboratory responses to coronavirus disease (COVID-19) in Africa. *Journal of Public Affairs*, *20*(4), 1–9. https://doi.org/10.1002/pa.2280.

Ogbo, A. (2015). *A phenomenological study of the influence of Nigerian leadership on the life of its citizens*. Dissertations & Theses @ University of Phoenix. https://www. proquest.com/dissertations-theses/phenomenological-study-influence-nigerian/docview/1711166292/se-2?accountid=35812.

Ogbo, A. (2016). Influence of leadership. *American Journal of Transformational Leadership*. Houston, TX.

Pollitt, C. (2016). The quest for good governance: How societies develop control of corruption. *International Review of Administrative Sciences*, *82*, 617–618.

Reuters.com. (2021). *Factbox: Africa's longest-serving leaders*. https://www.reuters.com/article/us-africa-presidents-factbox/factbox-africas-longest-serving-leaders-idUSKBN29I1N2

Shane, S. (2009). *Technology strategy for managers and entrepreneurs* (1st ed.). Upper Saddle River, NJ: Pearson.

Sharma, S., Kumari, B., Ali, A., Yadav, R., Sharma, A., Sharma, K., Hajela, K., & Singh, G. (2022). Mobile technology: A tool for healthcare and a boon in pandemic. *Journal of Family Medicine & Primary Care*, *11*(1), 37–43. https://doi.org/10.4103/jfmpc .jfmpc_1114_21.

Statista.com. (2022). *Number of coronavirus (COVID-19) deaths in the African continent as of November 22, 2021, by country*. https://www.statista.com/statistics/1170530/ coronavirus-deaths-in-Africa/

Techo, V. P. (2021). Optimistic leadership for the 21st century and beyond. *The Journal of Values-Based Leadership, 14*(2), 106.

The New York Times (January 31, 2022). *Five African Countries. Six Coups. Why Now?.* https://www.nytimes.com/article/burkina-faso-africa-coup.html

The UN Climate Change Conference COP 26. (2021). *Glasgow, U.K.* https://ukcop26. org/cop26-goals/.

Transparency International. (January 2022). *CPI 2021 for sub-Saharan Africa: Amid democratic turbulence, deep-seated corruption exacerbates threats to freedoms.* https://www.transparency.org/en/news/cpi-2021-sub-saharan-africa-amid-democratic-turbulence-deep-seated-corruption.

Transparency International. (2022). *Corruption Perceptions Index 2021.* https://images. transparencycdn.org/images/CPI2021_Report_EN-web.pdf.

UN Climate Change News. (2020, October 27). *Climate Change Is an Increasing Threat to Africa.* https://unfccc.int/news/climate-change-is-an-increasing-threat-to-africa.

Walter, J. (2007). Political leadership. In *Oxford Companion to Australian Politics.* Retrieved June 26, 2022, from https://www.oxfordreference.com/view/10.1093/acref /9780195555431.001.0001/acref-9780195555431-e-270.

World Bank. (2018). *Combating corruption.* http://www.worldbank.org/en/topic/governance/brief/anti-corruption.

World Meteorological Organization (WMO) (2020). State of the Climate in Africa 2019. Chair, Publications Board. CH-1211 Geneva 2, Switzerland.

Worldatlas.com (2021). *How Many Countries Are There In Africa?.* https://www.worldatlas.com/ articles/how-many-countries-are-in-africa.html.

Yukl, G. (2013). *Leadership in organizations* (8th ed.). Upper Saddle River, NJ: Prentice Hal

Chapter 13

Information Governance

A Catalyst for Digital Transformation

By Henrietta Okoro

INTRODUCTION

In today's regulated environment, the interconnection of systems and new technologies has contributed to risk management in digital transformation. Organizations are continually under competitive pressures and forced to re-evaluate their business models and underline business processes (Glavan & Kolic, 2021; Škrinjar et al., 2007). The digital transformation era is bringing together the three Ps, (people, data, and processes) that create technology disruption and transform global organizations to create value for the stake-holders and customers. Similarly, digital innovation has been a critical driver for developing societies and economies in the 21st century. According to the United Nations (2020), digital technologies have evolved faster than any other breakthrough in human history, reaching around 50% of the world's population in less than two decades (Hai & Tran, 2022).

Information Governance portrays a well-implemented information management program for organizations to manage data, have a strategic opportunity, analyze their data, and keep data safe, especially in the global environment. Information governance as a catalyst for digital transformation provides steps to aid effective strategy that quantifies the business and compliance risks specific within an organization. According to Drechsler et al. (2020), digital technologies not only enable unconstrained innovation but predominantly focus on a process called IT consumerization, which significantly alters the mindset and behaviors of workers and customers in the business environment (Gregory et al., 2018). This chapter discusses information governance and digital transformation from three corporate sectors: healthcare, telecommunication, and banking, and their compliance and regulations using COBIT, HIPAA, and ISO 27001.

GLOBAL AND ORGANIZATIONAL SECTORS PERSPECTIVES

Information Governance (IG) Overview

Information governance is the strategy for information in an organization. Information governance manages the risk and value that the report presents. Information governance helps in every aspect of the organization, such as legal compliance, operational transparency, budgeting, and other multitasking discoveries. Effective information governance begins with assessing how information is collected and how to store data, both digitally and non-digital. Information governance involves storing data and how an organization intends to use the data more, especially in a global network environment. According to (Coulson-Thomas, 2018), there is a decrease in information governance innovation despite the many possibilities of developing individual companies and their networks of relationships. Many boards are still reluctant to adopt new business, governance, and operation models in their organizations. They adhere to traditional practices and standard models often derived from companies that have declined in their legacy businesses. Thus, data sharing is vital, particularly with rules and regulations concerning the use of data becoming more robust. Information governance practices frequently include using information governance applications, which facilitate businesses in managing their data, availability of data, and regulatory compliance.

Digital Transformation Overview

Digital transformation is integrating digital knowledge and technologies into all business areas and changing how businesses operate and value their customers. Digital transformation also involves cultural change that requires organizations to continually challenge the status quo and possible research options and use failure as a learning process. New digital paradigms include a vast array of enabling digital technologies (DT), such as the Internet of things (IoT), additive manufacturing, big data, artificial intelligence, cloud computing, and virtual reality (Garzoni et al., 2020; Nambisan, 2017). Digital transformation strives to create new or modify existing functions and processes to meet dynamic business environments and market requirements. Thus, the reimagining of business in the digital age is digital transformation. The concept of digital transformation focuses on using technology to redesign processes for more efficiency. It is all about enhancement and changing the status quo for the better. Some examples of the technologies integrated into digital transformation projects include blockchain, big data, IoT, cloud computing, artificial intelligence (AI), and machine learning (Guan et al., 2022). In April

2021, the Flexjobs survey of over 1,100 working parents found that 62% of respondents said they "would quit their current job if they cannot continue remote work" (Reynolds, 2021). Thus, making digital transformation a must for every organization. Hence, digital transformation is more than just adding technology components to the transformation. Instead, it includes changing how employees work, think, and participate in projects. For instance, if an organizational culture does not promote change, it becomes challenging for the company to instill new business processes and attain digital insight. The business shift into a digital transformation often means breaking down silos and relating differently to employees, stakeholders, customers, and the public.

The Need for Digital Transformation in Various Sectors

In a competitive market, every organization, local and global, spikes inventive thinking. Digital transformation requires organizations to stay afloat by observing what is happening in their respective sectors. For example, having an online showroom to showcase companies' products relates to digital transformative techniques that allow customers to shop online, compare prices, and check for the best value and quality products. According to the N-iX survey, by 2023, the spending on the technologies and services that enable digital transformation worldwide is expected to amount to 2.3 trillion U.S. dollars (see Figure 1). Consequently, the benefits of change include

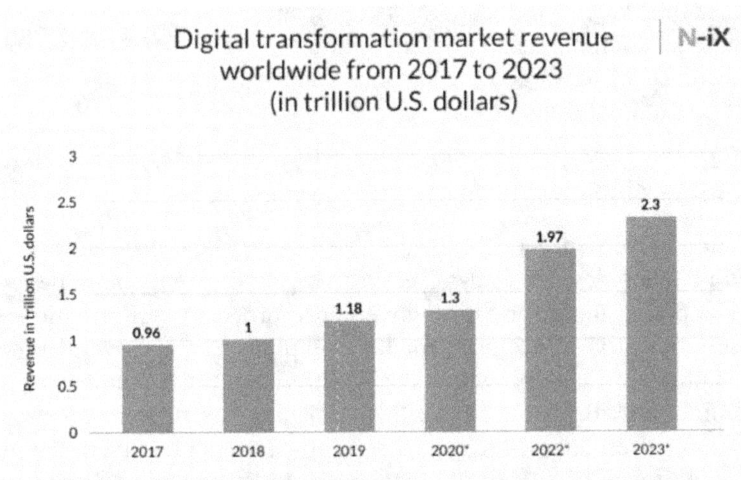

Figure 13.1 Digital Transformation Market Revenue Worldwide from 2017 to 2023. N-iX (November 10, 2020). Digital Transformation: Featured success stories. Retrieved from https://www.n-ix.com/digital-transformation-case-studies/

- To allow employees to integrate IT with the company's goals,
- To use technology to define processes clearly
- To enable consumerization of technology more ease of business to business (B2B) and more.
- To break down silos and deploy best-practice analytics across multiple sites and suppliers.
- To focus on improving working capital efficiency and on-time delivery performance
- The COVID-19 pandemic rapidly shaped digital transformation due to the shutdown of businesses and social distancing to eliminate the spreading of viruses.
- The ease of working from home or remotely to continue business operations during the pandemic.

Information Governance as a Mechanism to Digital Transformation

Information governance is a strategic approach to managing data and information in a digital and non-digital format to support business performance. IG involves a wide range of cross-disciplinary procedures, policies, controls, tools, and technologies that facilitates the company to meet operational demands and regulatory compliances. By integrating the proper use of data against regulatory demands, information governance helps to:

- Increase the value of that data to the company
- Allow legal compliance and risk mitigation
- Reduce costs of unwanted transactions and regulatory penalty
- Boost operational transparency

Information Governance in Healthcare Sectors

Information governance in the healthcare industry has positively impacted the sector's outcomes due to data security and privacy which are being managed by applying the information governance processes. Information governance sets objectives and goals for the compliance of the security policies in the health care industry, which deals with the massive number of clients' confidential data (Brown & Toze, 2017). Thus, it is undoubtedly said that information governance is among the significant technological actions that address data-related issues in the healthcare industry and all organizations. According to Smallwood (2018), information governance in the health care industry needs proper development of goals to impact health care decision-making processes positively and effectively. The author suggests that the healthcare organizations that have adopted the use of information governance are thus using the current technology and focusing on protecting the industry's information and, most importantly, clients. In addition, the author adds that information governance is crucial to the healthcare industry because it serves various purposes of ensuring that the healthcare workers use the compliance rules in protecting the information and managing the information systems well. Therefore, it mitigates the risks and threats that the healthcare industry comes across because the employees can handle the activities in a free-risk environment.

Global Impact on Healthcare

According to (Echendu and Claver, 2021), Africa is the fastest urbanizing continent. By 2050, the nation could count more than 2.5 billion people, with 55% of the population living in urban centers (Güneralp et al., 2017). The urban centers are central to achieving sustainable development because of the significant economic potential and opportunities. However, the vast amount of pollution stemming from cities from sources, among which transportation is an essential factor impacting sustainable development (Uduporuwa, 2020). African countries have weak healthcare infrastructure (Dekker et al., 2020; Fortune, 2021). Thus, the need to integrate information governance in Africa facilitates healthcare infrastructure towards digital transformation.

Information Governance in Telecommunication Sectors

Digital transformation has led to various technologies that increase data collection and management. Information governance ensures the company's information assets are made available to those who need them while ensuring compliance and streamlining management (Peng, 2018). The telecommunication industry recognizes the importance of protecting data privacy to maintain the trust of its customers. There are essential privacy laws that have been established, and the company needs to abide by them to enhance its information governance. Privacy protection is a significant focus for Verizon. Thus, there are established measures meant to protect data privacy and security by enhancing customer protection and ensuring compliance with privacy legislation. The growth of information technology has led to an increase in the data collected. Thus, technologies such as big data have increased the collection and storage of data. Verizon has established detailed privacy policies to enhance the security of data (Granneman, 2018). The complex privacy programs have led to establishment of an organization's overall structure the Chief Privacy Officer oversees. This position aims to review and discuss the privacy risks that face the organization.

No matter the size of any company, it must deal with vast information in its enterprise systems and outside of them. The added challenges of following government, financial, and industry regulations and dealing with a growing number of lawsuits are making more and more businesses embrace Information Governance. Proper information storage and retention and strict information security are the foundation for all Governance programs. However, even though these responsibilities may seem overwhelming at first, effective Information Governance can impact long-term impact on any organization (Brooks, 2019). It is problematic for businesses to get all their information in one place and control it to help them make profits. However, the benefits of information governance, such as sales and marketing, make it a priority for many companies today.

Information Governance in Banking Sectors

As the digital age approached, the first setback occurred down the road when the merger of Wells Fargo and Los Angeles bank, Interstate Bancorp, took place. The complex system integration went wrong, and the computer glitch resulted in many customers facing banking issues like losing their deposits and checks bouncing without reason. The trust and standard of Wells Fargo went down. As the world approached ever-growing digitization, the banking sector began changing, and the United States is at the center of the digital revolution. The banking industry had to adapt faster than ever in the United States. This paper's particular event that we are shedding light on happened in 2018. For information governance, the year 2018 is essential as European Union implemented General Data Protection Regulations (GDPR) and initiated an effort for data and information accountability and protection compliance all over the globe.

Digital financial services (DFS) are significant financial resolutions promoting financial inclusion (Buckley & Malady, 2015). DFS provides services to the underprivileged using advanced skills, digital platforms, and electronic money models (Scott et al., 2017). Digital channels can allow costs to be substantially low for both customers and service providers through isolated and underserved residents (Baker et al., 2007). Adoption and digital technologies enable millions of people to move out of poverty every year (Gates Foundation, 2017). However, a vast majority of the world's population falls into poverty due to various issues such as health, finance, and other astonishment—some people living in or near the poverty line lack even the most elementary banking services. (Rana et al., 2020).

Interestingly, in the year 2018, the Board of Directors of the Federal Reserve sent out letters to Wells Fargo company directors for an inefficient risk management portfolio. Wells Fargo was meant to follow SEC and

FINRA compliance and corporate governance rules as a publicly traded firm. Cross-selling products was the primary concern in banking since it included demanding targets and an incentive-based structure. The culture and pressure from higher management propelled Wells Fargo employees to create fraudulent sales and numbers (Witman, 2018). Governing bodies considered it a corporate governance flaw, but Corporate is a superset of information governance. The need for efficient information governance becomes imperative as the banking sector needs to comply with financial data and information compliances. SEC and FINRA are some preliminary bodies that provide governance regulations, including data and information storage regulations.

COMPLIANCE AND REGULATORY PERSPECTIVES

COBIT Overview

ISACA released the first version of COBIT in 1996, helping financial auditors improve controls within the IT environment. ISACA's second version of COBIT in 1998 shows additional rules beyond the audit community. Version 3 of the COBIT framework emerged in the early 2000s to include IT governance techniques and management guidelines that have become an integral part of the framework. The following versions — 4 (2005), 4.1 (2007), and 5 (2012) — provide substantial improvements in information regarding IT governance and risk management, especially in this pandemic era. The most recent version, COBIT 2019 (released in 2018), emphasizes the impact of information governance in driving organizational success. The latest version of COBIT provides users with flexible solutions that adapt to the rapidly changing technology of the modern IT landscape (Morteza & Masoud, 2020). COBIT principle 5, on separating management from governance: set clear boundaries between governance and management roles and responsibilities. Organizations achieve a holistic approach through the seven enablers: people, policies, frameworks, processes, organizational structures, culture, ethics, behavior, information, and services. Some of the importance of COBIT include:

- To maintain high-quality information to support business decisions.
- To achieve strategic goals and realize business benefits through effective and innovative use.
- To achieve operational excellence through reliable, efficient application of technology.

HIPAA Overview

The Health Insurance Portability and Accountability Act of 1996 (HIPAA) regulations were developed by the Secretary of the U.S. Department of Health and Human Services (HHS) to protect specific health information privacy and security. The Privacy Rule, or Standards for Privacy of Individually Identifiable Health Information, establishes national standards for protecting certain health information. The Security Standards for the Protection of Electronic Protected Health Information (the Security Rule) establish a national set of security standards for protecting certain health information held or transferred in electronic form (HHS.gov, n.d.). The healthcare industry expands beyond Health insurance Portability and Accountability Act 1996 (HIPAA) audit prep for Information Governance (IG). Healthcare organization has more privacy and security awareness efforts in a secure place. So, an actual key cybersecurity issue is to keep the information forever and not have any security problems with destroying data after that retention period has lapsed.

The Security Rule operationalizes the protections contained in the Privacy Rule by addressing the technical and non-technical safeguards that organizations called "covered entities" must put in place to secure individuals' "electronically protected health information" (e-PHI). Within HHS, the Office for Civil Rights (OCR) has responsibility for enforcing the Privacy and Security Rules with voluntary compliance activities and civil money penalties. Some of the importance of HIPAA includes

- To make electronic records safe and secure
- To simplify administrative processes
- To make insurance easy, portable, and accessible

The healthcare providers affected by HIPAA are physicians, doctors, dentists, chiropractors, and psychologists. Hence, the HIPPA Security Rule applies to health plans, health care clearinghouses, and any healthcare provider who transmits health information electronically in connection with a transaction for the "covered entities" and to their business associates (HHS .gov, n.d.).

ISO 27001

The ISO 27001 is also called ISO/IEC 27001 for Information technology, security techniques, and information security management systems requirements. ISO 27001 was published by the International Organization for Standardization (ISO) in partnership with the International Electrotechnical Commission (IEC). Both are leading international organizations that develop international standards. ISO-27001 is part of a set of measures designed to manage information security: the ISO/IEC 27000 series. (Culot et al., 2021). To regulate data protection and privacy in the European Union and the European Economic Area. The last ISO/IEC 27001 update in 2013 included the new regulatory requirements in the new ISO/IEC 27552 (Privacy Information Management) because of the need to ensure a secure environment for intellectual property to maintain attractiveness (Ku et al., 2009). Some of the importance of ISO 27001

- To identify stakeholders and their expectations of the company in terms of information security
- To identify what risks exist for the information
- To define controls (safeguards) and other mitigation methods to meet the identified expectations and handle risks
- To set clear objectives on what needs to be achieved with information security

DIGITAL TRANSFORMATION CHALLENGES

Digital transformation challenges for companies require more coordination through information governance when making decisions to accelerate the transformation process. Companies' information governance mechanisms around the impact of their digital initiatives include people, process, compliance, and technology. However, despite the digital transformation challenges, companies and stakeholders need to be abreast of the change as an opportunity to scale up the integration of information and communication technology (ICT) in business environments. Some of the inherent risks in digital transformation include:

- Inability to deploy solutions that can be adopted across all company levels.
- Inability to improve work processes
- Failure to target specific improvements.
- Teams focus mainly on technology instead of solving a vital business problem with a clear return on investment (ROI).

Response to Digital Transformation Challenges

To respond to digital challenges, organizations need to define effective Information governance. This Information governance driven by a competitive, technological, and artificial intelligence based on digital processes will penetrate beyond the automation and dematerialization projects of the information flows in the global terrain. A reflection on new products and services by digital technologies helps to maintain pace with competitors. In parallel, the marketing strategy using an effective information governance process will enhance the integration of digital marketing and transformation possibilities. The entire distribution channels in the digital transformation shall be analyzed to reach maximum clients. In the most extreme cases, a business strategy or model change must be considered to respond to digital transformation challenges.

Best Practices: Information Governance Mechanism to Digital Transformation

Information governance provides aids to sustain digital transformation into the 21st century. Today's business demand across every channel integrates online and offline transactions. Yet, companies lack ways to connect their customers to dynamic experiences with sales, service, and marketing. It discourages customers, causing them to switch companies after just one poor experience. The digital convergence of technology and behavior is only accelerating, and the effect it causes is transformative and disruptive. Markets are shifting to such an extent that they open the door to modernization, with new products, services, and ways of doing business becoming the norm.

Information governance with digital transformation guides one of today's most important movements in business evolution. Digital Transformation leads to opportunities to navigate the complexities of uncertainty, bureaucracy, politics, skepticism, fear, etc., towards progress. The factors for successful digital transformation include

Factors for Successful Digital Transformation

Early Orientation. The ability to create a new perspective to drive a meaningful change. The update focuses on customer needs, wants, and priorities when an organization is built around internal processes and functions.

Managing People. Understanding clients' and customer values, expectations, and behaviors have become imperative in the digital transformation era. Managing people requires top management to buy into the digital transformation and include the teams and stakeholders to align the entire organization with digital goals and strategies.

Policy and Processes. Constant organizational changes require assessment of operational infrastructure and the need to update technologies, processes, and policies to support change. Enhancing the career divisions is vital to ensure the delivery of outstanding customer experiences, team building, and collaborative opportunities.

Objectives and Strategies. Having a clear and concise purpose of digital transformation aligns with stakeholders around the new vision and strategies. Setting goals for the digital transformation to identify the species to improve. The key performance indicators (KPI) metrics are used to measure the performance and evaluate processes to help ensure that the company meets the digital transformation objectives.

Structure and Formation. Create a solid structure and form a dedicated digital experience team with clear responsibilities mapped out. Ensure that the entire team is aware of objectives and understands the processes to focus on the digital transformation purpose.

Insights and Intent. Identify, collect data, and apply insights to the strategy to manage digital evolution. Data could help streamline experiences across customer transactions and various brands. Data could also help in the evaluation of the results, brand relevance, increased revenue, and other processes in digital transformation.

Technology Integration. Technology helps to speed processes and delivers security for organizations. One needs an integrated approach to network security to move ahead with a broader mission and sustain competitive advantage. Advanced threat protection provides protections to ensure effective organizational critical data integration and management. In addition, end-to-end visibility allows companies to network and manage the entire system. Advanced analytics rapidly helps digital transformation from one network into another and provides insights for change management. The embracement of digital transformation allows the re-evaluation of front and back-end systems for seamless, integrated, and native customer experiences and, ultimately, employee experiences. Hence, using technology to promote trustworthiness and meet ever-increasing customer expectations becomes imperative.

Execution and Implementation. The process of implementing a strategic plan helps to put plans into action. The ability to learn and adapt to dynamic situation enable steering the ongoing digital transformation and customer experience workflow. The need to evaluate the state of the company's transformation to ensure that adjustments are made if necessary. When planning and implementing a digital transformation strategy, keeping an open mind is necessary to facilitate the organizational change process, monitoring, and readjustment.

Examples of Successful Digital Transformation Projects

Some successful companies that show best practices in digital transformation projects include Apple, BMW, and Amazon; with the financial industry changing, blockchain makes an impact, thus posing challenges to some sectors such as education and airline. Due to digital transformation, does

BMW change their car every two years instead of the earlier six years? Due to technological advancement, people need the latest and most significant innovations. For example, electronic vehicles (EV), car models: driverless cars, keyless, parking cameras, and other features to attract more customers. For example, Amazon uses algorithms for order fulfillment (Irani, 2015), and this creates work scenarios and orders that direct the work of workers in warehouses and dictate the speed of their work so that the worker can work efficiently and improve performance (Cesarovic et al., 2021).

Best Practices: Overcoming Digital Transformation Challenges

Among the best practices to help in overcoming digital transformation challenges include the following:

1. They have personalized learning tools to help people learn innovations and those with reading or writing challenges.
2. Digital transformation business context to grasp the attention of early starters in the business.
3. A change framework is the main factor in bringing clarity to a complex task
4. Integration of strategic and execution criteria
5. Assessing factors needed for managing digital transformation and driving change
6. Recruiting people with the required skills to execute and maintain the transformation.
7. Prioritize customer-facing applications to drive business value.
8. Align transformation efforts to expected business goals.
9. Allocate sufficient time (1 year to 2 years) to adjust to the digital transformation process.
10. Develop plans and a checklist for concrete corrective actions.

CONCLUSION

For companies looking to stimulate their digital transformation efforts, these factors provide a blueprint for stakeholders across the organization to come together, create a shared vision and take the first. Since patients' health information security and privacy are of great importance, it can be noted that. Therefore, ensuring that the safety information that has not yet been breached should be safeguarded by putting more stringent operations into practice to ensure it remains safe throughout the digital transformation process. Safety information helps to maintain the privacy of structured and unstructured data and harmonizes the team's responsibilities. Developing information governance systems in their workflow would respond to digital challenges positively, as most industries have developed modern computing technologies to address pertinent issues.

Furthermore, establishing a data protection cycle that covers all the steps in data collection, transmission, storage, processing, sharing, and destruction is essential in digital transformation. There are several ways to build up a defense system. Still, one way is to use technologies like data source verification, large-scale encryption, non-relational database encryption, data transaction security, data leakage prevention, and data destruction to make the system even more robust.

In addition, transformation becomes necessary to improve the security of big data platforms in the digital. User and component identity verification, fine-grained access control, a data operation security audit, and data desensitization will help to keep data from being accessed or leaked at the organizational level. Meanwhile, more attention should be paid to security issues when significant data platform components are set up and run. There should be an effective response to ensure proper management of the activities and enhance important data security in the digital transformation process. Thus, the information governance framework needs to be followed correctly in every sector within the global arena to facilitate digital transformation. Data protection and confidentiality are valuable aspects for every industry, and they can cause the downstream of an organization.

REFERENCES

Baker, E.W., Al-Gahtani, S.S. and Hubona, G.S. (2007). The effects of gender and age on new technology implementation in a developing country: testing the theory of planned behavior (TPB), Information Technology & People, Vol. 20 No. 4, pp. 352-375.

Brooks, J. (2019). Perspectives on the relationship between records management and information governance. Records Management Journal. https://www.emerald.com/insight/content/doi/10.1108/RMJ-09-2018-0032/full/html

Brown, D. C., & Toze, S. (2017). Information governance in digitized public administration. *Canadian public administration*, *60*(4), 581-604.

Buckley, R.P. (2015). Consumer protection issues for digital financial services, Experts Group on Financial Inclusion Policy (EGFIP) Meeting & Consumer Protection Regulatory Training Tonga, Nuku'alofa, Tonga, available at: www.clmr.unsw.edu.au/sites/default/files/attached_files/consumer_protection_issues_for_digital_financial_services_nov_18_2015_-_tonga_training_read-only.pdf

Cesarovic, A., Bogavac, M., & Cekerevac, Z. (2021). *The Impact Of The Pandemic On The E-commerce Of Small And Medium Enterprises*. Varazdin Development and Entrepreneurship Agency (VADEA).

Coulson-Thomas, C. (2021). Strategic thinking and corporate governance. *Effective Executive, 24*(4), 7-29. https://coloradotech.idm.oclc.org/login?url=https://www.proquest.com/scholarly-journals/strategic-thinking-corporate-governance/docview/2622692739/se-2?accountid=144789

Culot, G., Nassimbeni, G., Podrecca, M., & Sartor, M. (2021). The ISO/IEC 27001 information security management standard: Literature review and theory-based research agenda. TQM Journal, 33(7), 76-105. http://dx.doi.org/10.1108/TQM-09-2020-0202

Drechsler, K., Gregory, R., Heinz-Theo Wagner, & Tumbas, S. (2020). At the crossroads between digital innovation and digital transformation. *Communications of the Association for Information Systems, 47*, 23. http://dx.doi.org/10.17705/1CAIS.04723

Echendu, A. J., & Peter Claver, C. O. (2021). Smart city technology: A potential solution to Africa's growing population and rapid urbanization? *Development Studies Research, 8*(1), 82-93. http://dx.doi.org/10.1080/21665095.2021.1894963

Fortune, N. (2021). Digitalisation and sustainable energy transitions in Africa: Assessing the impact of policy and regulatory environments on the energy sector in Nigeria and South Africa. *Energy, Sustainability and Society, 11*(1)http://dx.doi.org/10.1186/s13705-021-00325-1

Garzoni, A., De Turi, I., Secundo, G., & Pasquale, D. V. (2020). Fostering digital transformation of SMEs: A four levels approach. *Management Decision, 58*(8), 1543-1562. http://dx.doi.org/10.1108/MD-07-2019-0939

Gates Foundation (2017), "Digital payments can benefit the poor and be good for business," A Report by Gates Foundation, available at: www.gatesfoundation.org/Media-Center/Press-Releases/2013/09/Digital-Payments-Can-Benefit-the-Poor

Glavan, L. M., & Kolic, I. (2021). *Transformation Of Business Process Management Through Digitalization*. Varazdin Development and Entrepreneurship Agency (VADEA).

Gregory, R. W., Kaganer, E., Henfridsson, O., & Ruch, T. J. (2018). IT consumerization and the transformation of IT governance. MIS Quarterly, 42(4), 1225-1253.

Guan, A. L. C., Manavalan, M., Ahmed, A. A. A., Azad, M. M., & Miah, M. S. (2022). Role of Internet of Things (IOT) in Enabling Productive Work from Home (WFH) for Environmental Volatiles. *Academy of Marketing Studies Journal, 26*(1), 1-11. https://coloradotech.idm.oclc.org/login?url=https://www.proquest.com/scholarly-journals/role-internet-things-iot-enabling-productive-work/docview/2618819397/se-2?accountid=144789

Güneralp, B., S. Lwasa, H. Masundire, S. Parnell, and Seto, K. C. (2017). Urbanization in Africa: Challenges and Opportunities for Conservation. Environmental Research Letters 13 (1): 015002.

Hai, N. N., & Tran, M. D. (2022). Stimuli to adopt e-government services during covid-19: Evidence from Vietnam. *Innovative Marketing, 18*(1), 12-22. http://dx.doi.org/10.21511/im.18(1).2022.02

HHS.gov (n.d.) U.S. Department of Health & Human Services. Retrieved from https://www.hhs.gov/hipaa/for-professionals/security/laws-regulations/index.html

Morteza, A., & Mir-Amini Masoud. (2020). Integrating COBIT with a hybrid group decision-making approach for a business-aligned IT roadmap formulation. Information Technology and Management, 21(2), 63-94. http://dx.doi.org/10.1007/s10799-019-00305-0

Nambisan, S. (2017). Digital entrepreneurship: toward a digital technology perspective of entrepreneurship, Entrepreneurship: Theory and Practice, Vol. 41 No. 6, pp. 1029-1055.

Rana, N. P., Luthra, S., & Rao, H. R. (2020). Key challenges to digital financial services in emerging economies: The Indian context. *Information Technology & People, 33*(1), 198-229. http://dx.doi.org/10.1108/ITP-05-2018-0243

Reynolds B. W. (2021). FlexJobs Survey: Working Parents Want Remote Work. Retrieved from https://www.flexjobs.com/blog/post/what-working-parents-want-at-work

Scott, S.V., Van Reenen, J. and Zachariadis, M. (2017), "The long-term effect of digital innovation on bank performance: an empirical study of SWIFT adoption in financial services", Research Policy, Vol. 46 No. 5, pp. 984-1004.

Škrinjar, R., Indihar-Štemberger, M. and Hernaus, T. (2007), The Impact of Business Process Orientation on Organizational Performance, Proceedings of the 2007 Informing Science and IT Education Joint Conference, Ljubljana, 171-185.

Smallwood, R. F. (2018). *Information governance for healthcare professionals: a practical approach.* Productivity Press.

Uduporuwa, R. (2020). Sustainable City Development Is Possible? A Review of Challenges and Key Practices Towards Urban Development in Developing Countries. International Journal of Scientific and Research Publications (IJSRP) 10 (9): 294–304.

United Nations. (2020). The Impact of Digital Technologies. Retrieved from https://www.un.org/en/un75/impactdigital-technologies

Witman, P. D. (2018). "What gets measured, gets managed" the Wells Fargo account opening scandal. Journal of Information Systems Education, 29(3), 131-138. Retrieved from https://www.proquest.com

Index

About the Authors

Kofo Aderogba has extensive background in Geography and Environmental Management, a Professor in the Department of Adult and Development Education, Tai Solarin University of Education, Ijagun, Nigeria. Her research efforts have yielded 6 edited books, over 80 articles and book chapters, and presentations more than 45 in local and international conferences and workshops. She has widely travelled. Professor Aderogba is the alternate chairperson on the Board of Greenland Hall Group of Schools, and the chairperson on the Board of Hand of Mercy Foundation

Syed Adeel Ahmed holds a PhD in Engineering and Applied Sciences, two Master of Science degrees in Electrical Engineering (MSEE) and Engineering Management (MSENMG) and Bachelor of Science in Electronics and Communication Engineering. He is a Microsoft Certified Professional and Business Strategy Game Champion. Dr. Ahmed is also Faculty at Tulane University, the University of New Orleans, Xavier University, Southern University of New Orleans, Dillard University, Delgado Community College, and Nunez Community College. He serves on Climate Reality, Interfaith and CLEP research boards and a Member of the World Association Academic Doctors. He may be reached at sahmed1@xula.edu.

Bimpe E. Alabi is a graduate of Adult Education from the Department of Adult and Lifelong Learning, Obafemi Awolowo University, Ile-Ife, Nigeria. She is currently a lecturer in the Department of Adult and Development Education, College of Specialized and Professional Education, Tai Solarin University of Education, Ijagun, Nigeria. She has authored chapters in books and contributed articles in highly referred professional journals. Her research interests and community service engagements focus on human rights, women, children, and community development.

Theophilus Boufini holds a PhD in Finance and fellowship of both the Institute of Chartered Accountants of Nigeria, (ICAN), Institute of Corporate Administrators of Nigeria and membership of the Project Management Institute, (PMI), USA. He is a Management and Financial Consultant, Trainer and Motivator specializing in Human Resources, Management, Leadership, Organizational behavior, Financial re-engineering, Budget and budgetary control and project management with over 25 years working experience which cuts across the print media, banking and the upstream oil and gas industry.

Gina Brown currently serves as the Dean for the College of Nursing and Allied Health Sciences at Howard University in Washington, DC. For over 25 years. Dr. Brown has worked professionally in North America, Asia, and Africa. Her expertise as a professional administrator (educator and nurse), in addition to her wide academic experience in many areas (traditional, adult and on-line learning) has allowed for the development of graduate and undergraduate programs.

Jessica L. Burke is an Associate Professor of Sociology at Francis Marion University. Dr. Burke earned her PhD in Social Psychology and Mental Health, MSc in Health, and Healthcare. Her research focuses on issues related to health, including distress and domestic violence, and health outcomes related to domestic violence during periods of social crises. Dr. Burke has presented at several national, regional, and state conferences, including the Society for the Study of Social Problems, the Society for the Study of Symbolic Interaction, the Southern Sociological Society, and the American Society of Criminology.

Lisa A. Eargle is a Professor of Sociology and Board of Trustees Research Scholar at Francis Marion University in Florence, S.C. Her teaching and research focus primarily on the areas of environment, disasters, and crime. Dr. Eargle received a McNair Center grant to study the impacts of Hurricane Matthew on the resources and well-being of the residents in the Pee Dee region of S.C. Her work appears in numerous other edited books and peer-reviewed journals. Her email address is leargle@fmarion.edu

Sheri Harshaw is a native of Detroit, Michigan. She holds a PhD in Business Management with a concentration in Organizational Management, a master's degree in Higher Education Administration and Leadership and a Bachelor of Science in Biology. She has a certification in Environmental Engineer from

Lawrence Technological College in Detroit, Michigan. Dr. Harshaw is the CEO and President of Female Managers/Supervisors (FMS) Management Service a support service that prides itself on the enrichment of women in management positions striving to move higher in their careers through inclusive programs designed by women and for women.

Tolu O Isele holds master's degrees in E-commerce Law and Oil & Gas Management and LLB from Madonna University, Okija, Anambra State, Nigeria. She is Head of Legal Unit at Greenland Hall Group of Schools, Dalemo, Alakuko, Ogun State, Nigeria. Ms. Isele is on the Board of Greenland Hall Group of Schools. She is versatile in school administration, and passionate about human rights and community development.

Compreca Martin received her PhD in Health Administration from University of Phoenix in 2015. She teaches for multiple universities and also serves as a Doctoral Committee Chair for Walden University. Outside of her passion for teaching, Dr. Martin is learning Mandarin Chinese. Her personal research interests are related to vulnerable populations, violence prevention, and community action-research.

Dana Mashal was Dr. Okunji's graduate student in the Family Nurse Practitioner program at the Department of Nursing when the book chapter was initiated. She is currently a clinical nurse at Howard University Hospital and George Mason University.

Brendan James Moore is a philosopher and instructional designer currently working on a leadership development program at Ochsner Health Systems in New Orleans, Louisiana. His background includes 10+ years of university ethics teaching at Ohio University and Tulane University and several years of work in the area of information technology, instructional technology, and applied computing systems. He is a PhD student in engineering and is passionate about teaching, process improvement, energy efficiency, and philosophy. He may be reached at bmoore@tulane.edu.

Anthony Obi Ogbo is the publisher of Texas International Guardian and an adjunct Professor of Communication and Journalism with Texas Southern University, Houston. He is a member of the Editorial Board of *The West African Pilot News*, a multimedia entity he co-founded in January 2020. An accomplished author of numerous books and journals, Dr. Ogbo's research interests focus on the fields of management and leadership as they apply to all academic disciplines, including communication, human capital, business administration, organizational leadership, health, applied econometrics, and development economics.

Priscilla Okunji is an Associate Professor at Howard University, College of Nursing and Allied Health Sciences. She obtained her PhD in Health Sciences with specialization in International Health Education, MSc, and BSc in Nursing Informatics from the University of Maryland. She also has degrees in Medical Microbiology, a post-graduate Diploma in Education and a baccalaureate degree in Microbiology were all obtained from the University of Nigeria. Her professional background merges over thirty-four years of teaching, research and clinical experience in Education, Microbiology, Nursing, and Informatics.

Twila Sterling-Guillory PhD, APRN, FNP-BC received a BSc and MSc in Nursing and a PhD in Nursing Research. She has practiced as a Nurse Practitioner in several different clinical settings. She is currently practicing as a Nurse Practitioner in the hospice setting. She was promoted to the rank of Professor in the College of Nursing and Health Professions in Spring of 2021 at McNeese State University in Lake Charles, Louisiana. She was the first African American faculty member in the College of Nursing and Health Professions to be promoted to this rank in the 65-year history of the College. She is a member of Sigma Theta Tau International and Alpha Kappa Alpha Sorority Inc.

Caron Strong is the Co-Chair of the Graduate Nursing at Howard University College of Nursing and Allied Health Sciences Graduate Nursing Program. Dr. Strong is also a Family Nurse Practitioner with clinical experience in family practice. Dr. Strong worked as a pediatric nurse. Dr. Strong has conducted and published research on Depression in the Primary Care Setting. Her academic interests are online teaching and using technology in the classroom.

Ghosh Thinnakkakath has many years of IT consulting and management experience at various organizations. Besides his MS in mathematics, Ghosh holds a Post Graduate Diploma in Computer Applications, an MBA, and a PhD in Information Technology. Member of the World Association of Academic Doctors (waadinc.org) and PMP certified (Project Management Institute). As a lifelong learner, Ghosh believes in servant leadership and mentoring/training others to raise them to the next level.

Devora Winkfield serves as Director of Graduate & Interim Chair of Undergraduate Nursing at Howard University. She is committed to the education of the next generation of health care providers. Dr. Winkfield intermittently practices in the Washington, DC area in community health clinics that serve the under-insured population. Over the past 20 years, she has served as a nursing instructor, professor, and department chair at the undergraduate and graduate levels in various institutions of higher learning.

Veronica D Woodard is a native of Eunice, Louisiana and a retired Associate Professor who taught at McNeese State University in the College of Nursing and Health Professions for 16 years. She has amassed an impressive 44 years in the profession of nursing. Her experience has covered different facets of nursing including, Clinical Nurse Specialist, entrepreneur (home health agency owner for 10 years), and educator. She holds degrees from Dillard University, LSU Medical Center Graduate School of Nursing, (currently LSU Health Sciences Center), New Orleans, and Duplichain University.

About the Editors

Ashraf Esmail is an Associate Professor and Program Coordinator of Criminal Justice at Dillard University and serves as the Director for the Center for Racial Justice. He has a Ph.D. in Interdisciplinary Studies with focus in Sociology and Educational Leadership and Administration and a minor in Criminal Justice, MA in Sociology, MS in Criminal Justice, and BA in Sociology. He currently serves as Senior Editor for the *Journal of Education and Social Justice,* Chief Editor for the *International Journal of Leadership, Education and Business Studies* and Co-editor of the *Journal of Contemporary Issues in Higher Education.*

Brandon Hamann is an Adjunct Professor of Criminal Justice at Dillard University of New Orleans. He is a Research Criminologist with a master's degree in Criminology and Justice from Loyola University of New Orleans and is the Co-Editor for the *Journal of Education and Social Justice,* a member of the Editorial Board for the *International Journal of Leadership, Education and Business Studies,* as well as a freelance book editor. Brandon is credited with having works in publications ranging in topics from Police Brutality, Domestic Terrorism, Cyber Warfare, and Climate Change.

M. Henrietta Okoro is an Adjunct Professor of Information Technology and Business Programs at Colorado Technical University, National University, and the University of the Cumberland. She is the founder/CEO of two nonprofit organizations: the World Association for Academic Doctors (WAAD) and the International Association for Female Academic Doctors (IAFAD). Dr. Okoro holds a Doctor of Management in Organizational Leadership specializing in Information Systems and Technology, a dual Master of Business Administration degree in Accounting and Management, and a Bachelor of

Arts degree in Business Administration. She is an alumnus of Harvard University, John F. Kennedy School of Government, Cambridge, Massachusetts, USA, where she completed Executive Education in Leadership & Decision Making.

Edward Perekebina Agbai is an author and management professional with backgrounds in Leadership, Research, Change Management, and Engineering. He is the Secretary of the World Association of Academic Doctors. Dr. Agbai holds a PhD in Management with specialization in Leadership and Organizational Change, MSc in Project Management, and HND in Electrical Engineering. He started his career in 2000 with Eni, a major oil and gas player in Nigeria before joining Smart Alliance Consulting Services Ltd as Resource Partner. He is an Adjunct Faculty with Emmanuel University, Fayetteville, NC, USA.

Sylvester Alor Favor Udeorah is a development economist with special interest in Labor/Human Resources Economics. He currently works at the Department of Economics, Faculty of Social Sciences, University of Port Harcourt, Nigeria where he has been engaged in teaching, research, and community service. He has over forty publications; including articles in learned journals and chapters in books to his credit. Dr. Udeorah enjoys writing and nurturing ideas, and he is happily married with three children.

www.ingramcontent.com/pod-product-compliance
Lightning Source LLC
LaVergne TN
LVHW011808200926
848484LV00005B/1108
9780761873884